Complete
ICT
for IGCSE®

Stephen Doyle

OXFORD
UNIVERSITY PRESS

OXFORD
UNIVERSITY PRESS

Great Clarendon Street, Oxford OX2 6DP

Oxford University Press is a department of the University of Oxford.
It furthers the University's objective of excellence in research, scholarship,
and education by publishing worldwide in

Oxford New York

Auckland Cape Town Dar es Salaam Hong Kong Karachi
Kuala Lumpur Madrid Melbourne Mexico City Nairobi
New Delhi Shanghai Taipei Toronto

With offices in

Argentina Austria Brazil Chile Czech Republic France Greece
Guatemala Hungary Italy Japan Poland Portugal Singapore
South Korea Switzerland Thailand Turkey Ukraine Vietnam

© Oxford University Press 2012

British Library Cataloguing in Publication Data

Data available

ISBN 978 0 19 912906 5

10 9 8 7 6 5 4

Printed in China by Printplus

Paper used in the production of this book is a natural, recyclable product made from wood grown in
sustainable forests. The manufacturing process conforms to the environmental regulations of the country
of origin.

Acknowledgements

Cover image: Pasieka/Science Photo Library

p.1:© Timur Anikin/Fotolia; p.1:© Rob Byron/Fotolia; p.2:© Mike Kiev/Fotolia; p.2:© megasquib/Fotolia;
p.3:© Harris Shiffman/Fotolia; p.4:© BlueMiniu/Fotolia; p.5:© Kheng Guan Toh/Fotolia; p.5:© Sean Gladwell/
Fotolia; p.6:© Petr Ivanov/Fotolia; p.6:© Semenov Gleb/Fotolia; p.7:© Rafa Irusta/Fotolia; p.7:
© John Tomaselli/Fotolia; p.8:© norhazlan/Fotolia; p.8:© arrow/Fotolia; p.9:© Robert Lehmann/Fotolia;
p.11:© Michael Brown/Fotolia; p.12:© Dmitry Terentjev/Fotolia; p.12:© Inclusive Technology; p.12:© Red
Rice Media/Fotolia; p.13:© S.White/Fotolia; p.13:© Andres Rodrigo Gonzalez Buzzio/Fotolia; p.13:© Marek
Tihelka/Fotolia; p.13:© GLUE STOCK/Fotolia; p.14:© gemenacom/Fotolia; p.14:© hauhu/Fotolia; p.14:
© Alan Stockdale/Fotolia; p.15:© Krzysiek z Poczty/Fotolia; p.15:© Glasbergen; p.16:© Ronald V/Fotolia;
p.17:© Ewa Walicka/Fotolia; p.17:© Stephen VanHorn/Fotolia; p.18:© Kurt De Bruyn/Fotolia; p.18:
© Stephen Coburn/Fotolia; p.19:© IKO/Fotolia; p.19:© Glenda Powers/Fotolia; p.20:© Anatoly Vartanov/
Fotolia; p.21:© gogik/Fotolia; p.21:© Dmitry Goygel-Sokol/Fotolia; p.22:© JackF/Fotolia; p.22:© ussatlantis/
Fotolia; p.22:© rlat/Fotolia; p.23:© artSILENSEcom/Fotolia; p.23:© Graça Victoria/Fotolia; p.24:© Donald
Swartz/Fotolia; p.24:© Alx/Fotolia; p.24:© Stephen Coburn/Fotolia; p.24:© picsfi ve/Fotolia; p.24:© Anatoly
Vartanov/Fotolia; p.29:© skaljac/Fotolia; p.29:© Nikon'as/Fotolia; p.30:© c/Fotolia; p.30:© Yong Hian Lim/
Fotolia; p.30:© Nikolai Sorokin/Fotolia; p.31:© TAlexTech/Fotolia; p.32:© Kirsty Pargeter/Fotolia; p.32:
© Nikolai Sorokin/Fotolia; p.39:© Alex White/Fotolia; p.36:© Dark Vectorangel/Fotolia; p.37:© Jiri Hera;
p.42:© Konstantin Shevtsov/Fotolia; p.35:© titimel35/Fotolia; p.40:© Valerie potapova/Fotolia; p.41:
© spanky1/Fotolia; p.41:© Sean Gladwell/Fotolia; p.41:© Glasbergen; p.39:© Peter Galbraith/Fotolia; p.50:
© chrisharvey/Fotolia; p.53:© Martina Taylor/Fotolia; p.54:© Pixel/Fotolia; p.54:© Stephen Coburn/Fotolia;
p.54:© Glasbergen; p.55:© Rainer Plendl/Fotolia; p.58:© iQoncept/Fotolia; p.59:© beaucroft/Fotolia; p.60:
© Gina Sanders/Fotolia; p.60:© Glasbergen; p.61:© daseaford/Fotolia; p.61:© Monkey Business/Fotolia; p.66:©
streetphotoru/Fotolia; p.66:© vicky/Fotolia; p.69:© robynmac/Fotolia; p.70:© Stephen Finn/Fotolia; p.71:©
kapp/Fotolia; p.72:© b.neeser/Fotolia; p.72:© Pierre-Emmanuel Turcotte/Istock; p.72:
© www.hortibot.dk; p.72:© Baloncici/Shutterstock; p.72:© www.mowbot.co.uk; p.73:© Yakov Stavchansky/
Fotolia; p.73:© Zoe/Fotolia; p.76:© Jasmin Merdan/Fotolia; p.76:© Capita SIMS; p.76:© Dawn Hudson/
Fotolia; p.77:© palms/Fotolia; p.78:© OneO2/Fotolia; p.78:© patrimonio designs/Fotolia; p.79:© DX/Fotolia;
p.80:© Denis Pepin/Fotolia; p.80:© TebNad/Fotolia; p.81:© Dmitrij Yakovlev/Fotolia; p.81:© Retail Systems
Technology; p.86:© Arrow Studio/Fotolia; p.96: © Brilt/Fotolia; p.99:© Guy Erwood/Fotolia; p.106:© booka/
Fotolia; p.107:© Onidji/Fotolia; p.108:© Onidji/Fotolia; p.109:© ktsdesign/Fotolia; p.110:© SETI Institute;
p.232:© Stephen Doyle; p.263:© Stephen Doyle

We have tried to trace and contact all copyright holders before publication. If notified the publishers will be
pleased to rectify any errors or omissions at the earliest opportunity.

The author and publisher would like to thank Michael Gatens for his work as a consultant editor on this
project.

®IGCSE is the registered trademark of Cambridge International Examinations.

All past paper questions are reproduced by permission of Cambridge International Examinations. Cambridge
International Examinations bears no responsibility for the example answers to questions taken from its past
question papers which are contained in this publication.

Contents

Introduction

What the book covers

This book covers the material needed for the three papers for the IGCSE® Information and Communication Technology. The CD includes additional material which supplements the material in the student book.

How you will be assessed

The assessment for the IGCSE consists of the following three papers:

Paper 1

This is a written paper of two hours duration which tests sections 1–8 of the curriculum content.

The questions are all compulsory and many of them consist of multiple choice or short answer questions. There are others which require longer answers.

The marks for paper 1 are 40% of the total.

Paper 2

This is a practical test that assesses the knowledge, skills and understanding of sections 9–16 of the curriculum content.

The marks for paper 2 are 30% of the total.

Paper 3

This is a practical test that assesses the knowledge, skills and understanding of sections 9–16 of the curriculum content.

The marks for paper 3 are 30% of the total.

Sections forming the curriculum content

The curriculum content is divided into the following eight interreleated sections:

1 Types and components of computer systems
2 Input and output devices
3 Storage devices and media
4 Computer networks
5 Data types
6 The effects of using ICT
7 The ways in which ICT is used
8 Systems analysis and design

Assessment criteria for the practical tests

The assessment criteria for the practical tests are set out in the following eight sections:

9 Communication
10 Document production
11 Data manipulation
12 Integration
13 Output data
14 Data analysis
15 Website authoring
16 Presentation authoring

When your work is marked you will have to meet a series of learning outcomes in each of the sections outlined above.

For access to all the files used in the activities go to **www.oxfordsecondary.co.uk/completeict**, or the *Complete ICT for IGCSE Teacher Kit* CD-ROM.

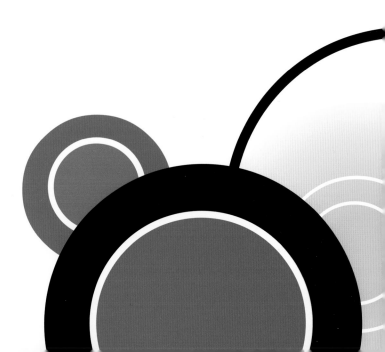

1 Types and components of computer systems

Computer and ICT systems used to be confined to desks but many people now want to use computers on the move. The size of a typical computer has reduced over the years and they have also become much easier to use. This first chapter looks at the differences between hardware and software and looks at the main components of a general-purpose computer. You will be looking at the different types of computers and also the recent developments in technology.

The key concepts covered in this chapter are:
▶▶ Definitions of hardware and software
▶▶ Differences between hardware and software
▶▶ The main components of a general-purpose computer
▶▶ Types of operating system
▶▶ Different types of computer (e.g., personal computer, mainframe, etc.)
▶▶ Recent developments in ICT

Hardware and software and the differences between them

All ICT systems consist of two main parts: the hardware and the software. Hardware means those parts of the computer you can touch. Here is a guide to help you decide if it is hardware or software.

▶▶ Is it classed as a computer or part of an ICT system? YES
▶▶ Can you touch it? YES
    Then it is **HARDWARE**

▶▶ Is it classed as part of an ICT system? YES
▶▶ Can you touch it? NO
    Then it is **SOFTWARE**

Hardware means any item of an ICT system that you can touch. This means that they are components such as printers, keyboards, memory, storage devices, etc. Even the storage media is classed as hardware. This means that a CD is hardware but if it has a program on it then the program itself is classed as software. It is hard to separate the program (i.e., the software) from the hardware that is used to store it.

Software means the programs that supply the instructions to the hardware to tell it what to do.

Examples of computer hardware

Hardware consists of the physical components of an ICT system and would include items such as:

▶▶ keyboard
▶▶ monitor
▶▶ processor
▶▶ speakers
▶▶ mouse
▶▶ fixed hard disks
▶▶ ROM/RAM
▶▶ printer
▶▶ web cam.

⊙ KEY WORDS

Hardware the parts of the computer that you can touch and handle.
Software the actual programs consisting of instructions that allow the hardware to do a useful job.

The processor is below the fan and the heat sink which removes the heat generated when the processor is working.

Storage (e.g., hard drive).

1

Keyboard.

Mouse.

Examples of computer software

The software is the set of instructions that tells the computer hardware what to do. It is written in a computer language and there are quite a few different ones. Computer hardware is useless without software, which is of two main types: operating systems and applications software.

Here are some examples of software:

▸▸ operating system software (e.g., Windows 7, Linux, Mac OS)
▸▸ word-processing package
▸▸ presentation package
▸▸ spreadsheet package
▸▸ web browser
▸▸ database
▸▸ CAD (computer-aided design)
▸▸ web design package
▸▸ photo editing package.

It is important to remember that the disks that software is installed on are classed as hardware. The programs themselves are the software.

! Revision Tip

When asked to name types of software, do not use brand names. So, for example, you should use "word-processing" rather than Microsoft Word in an answer.

Activity 1.1

How many pieces of hardware can you name?
There many different pieces of hardware that make up or can be used with computer systems. How many of them can you name? Produce a list and it is ok to use the names of the examples given in this chapter. About 20 different names would be a good total.

Activity 1.2

Hardware or software
Think about the differences between hardware and software and classify each of the following by putting a tick in the relevant box.

Name of item	Hardware	Software
Keyboard		
Operating system		
Scanner		
Remote control		
Word-processor		
Web browser		
Mouse		
Spreadsheet		
Database		
DVD		
Laser printer		
Virus checker		
CD ROM		
Web design		

Computer hardware

The main components of a general-purpose computer

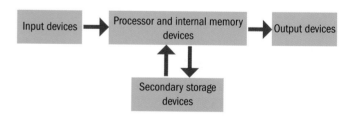

The main components of a general-purpose computer system.

The main components of a general-purpose computer are:

▸▸ Input devices (keyboard, mouse, etc.)
▸▸ Output devices (printer, monitor, speakers, etc.)
▸▸ Secondary storage devices (DVD-R/W drive, portable hard disk drive, etc.)
▸▸ Processor and internal memory devices.

The processor and internal memory devices include the central processing unit (CPU), read only memory (ROM), random access memory (RAM) and the internal hard disk drive.

Central processing unit (CPU)

The central processing unit (CPU), often called the processor, is the brain of the computer and it consists of millions of tiny circuits on a silicon chip. The central processing unit does a number of tasks: it controls the step-by-step running of the computer system, it does all the calculations and performs all the logical operations and deals with the storage of data and programs in memory.

Internal memory (i.e., ROM and RAM)

There are two types of memory called ROM and RAM. Both these two types of memory are stored on chips and are available immediately to the central processing unit (CPU). Memory (i.e., ROM and RAM) is often called primary storage.

Computers also have a hard disk drive as internal memory and it is here that the application software is stored along with the user's files.

ROM (read only memory)

ROM is fast permanent memory used for holding instructions needed to start up the computer.

ROM is:

▶▶ held on a computer chip
▶▶ called non-volatile memory because it does not lose its contents when the power is turned off
▶▶ used to hold instructions to start the computer, which are called the boot program or BIOS (basic input/output system). This finds out which devices are connected to the computer when it is first switched on and also loads the operating system
▶▶ used to store data that cannot be altered by the user.

RAM (random access memory)

RAM is a fast temporary memory where programs and data are stored only when the power is supplied.

RAM:

▶▶ is held on a computer chip.
▶▶ is called volatile memory because the contents disappear when the power is turned off.
▶▶ can be written to and read from.
▶▶ can be altered by the user.
▶▶ holds the software currently in use by the user.

The importance of ROM and RAM

Having a large amount of ROM and RAM in a computer is important, because having more memory means:

▶▶ applications run faster on their own
▶▶ more applications are able to run at the same time
▶▶ users are able to move quickly between applications.

RAM chips.

> **◉ KEY WORDS**
>
> **BIOS** (basic input/output system) stored in ROM and holds instructions used to "boot" (i.e., start) the computer up when first switched on.
> **RAM** random access memory – fast temporary memory which loses its contents when the power is turned off.
> **ROM** read only memory – memory stored on a chip which does not lose data when the power is turned off.

Input devices

These are used to get raw data into the computer ready for processing by the CPU. Some input devices, such as a mouse, keyboard, touch screen, microphone, etc., are manual and need to be operated by a human. Others are automatic and once they are set up they can be left to input the data on their own. These include optical mark readers, optical character readers, etc.

See Chapter 2 for more information on input.

Output devices

Once the raw data has been processed it becomes information and this information needs to be output from the computer using an output device. Output devices include monitors/screens, printers, speakers, plotters, etc.

See Chapter 2 for more information on output.

Backing storage

Secondary storage devices use removable media. Secondary/backing storage is used for the storage of programs and data that are not needed instantly by the computer. It is also used for long-term storage of programs and data as well as for backup copies in case the original data is lost.

3

Secondary/backing storage media includes portable hard disks, magnetic tape, memory sticks, flash memory cards, and optical disks such as CD and DVD.

See Chapter 2 for more information on storage devices.

QUESTIONS A

1 a Explain the main difference between computer hardware and software. *(2 marks)*

b Give **two** examples of computer hardware and **two** examples of computer software. *(4 marks)*

2 The diagram below represents the main components of a general-purpose computer:

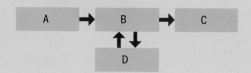

Here is a list of the components: Processor and internal memory devices, output devices, secondary storage devices, input devices.

Put the name of the device next to the correct letter. *(4 marks)*

A _____   C _____

B _____   D _____

3 ROM and RAM are both types of computer memory.

a i What do the letters ROM stand for? *(1 mark)*

ii What do the letters RAM stand for? *(1 mark)*

b Tick **one** box next to each item in the table to show which statements apply to ROM and which to RAM. *(4 marks)*

	ROM	RAM
Contents are lost when the computer is turned off	☐	☐
Contents are not lost when the computer is turned off	☐	☐
Stores the programs needed to start up the computer	☐	☐
Stores application programs and data currently being used	☐	☐

4 In computers and mobile devices such as mobile phones, palmtops, and PDAs, memory is needed.

a Give the full names of ROM and RAM. *(2 marks)*

b Explain why both ROM and RAM are required. *(2 marks)*

5 a Explain why it is important that a computer has a large amount of RAM and ROM. *(2 marks)*

b Computers contain RAM and ROM.
Explain what ROM would be used for in a computer.
Explain what RAM would be used for in a computer. *(4 marks)*

Operating systems

Operating systems are programs that control the hardware directly by giving the step-by-step instructions that tell the computer hardware what to do. An operating systems performs the following tasks:

▸▸ Handles inputs and outputs – selects and controls the operation of hardware devices such as keyboards, mice, scanners, printers, etc.

▸▸ Recognizes hardware – the operating system will recognize that a hardware devices such as a pen drive, camera, portable hard drive, etc., has been attached to the computer. The operating system loads the software it needs to control the device automatically.

▸▸ Supervises the running of other programs – it provides a way for applications software (i.e., the software that is used to complete a task such as word-processing, spreadsheet, stock control, etc.) to work with the hardware.

▸▸ Handles the storage of data – it keeps track of all the files and directories/folders on the disk.

▸▸ Maximizes the use of computer memory – the operating system decides where in the memory the program instructions are placed. For example, some instructions are needed over and over again, whereas others are only needed now and again. It ensures that the parts of the program needed frequently are put in the fastest part of the memory.

▸▸ Handles interrupts and decides what action to take – when something happens such as the printer cannot print because the paper is jammed or it has run out of paper, it will stop the printer and alert the user.

An operating system is software that instructs the hardware what to do.

The types of interface used with operating systems

An operating system needs a way of interacting with the user. The way the operating system communicates with a user is called the interface and there are two common interfaces in use and these are:

▶▶ Graphical user interface (GUI) – these are very easy to use and have features such as windows, icons, menus, pointers, etc. Examples of operating systems which make use of a GUI are:

   ▶▶ Windows

   ▶▶ Mac OS.

▶▶ Command line/driven interface – here you have to type in a series of commands. These commands have to be precisely worded and it can be hard to remember how to do this, so this type of interface is harder to use. An example of an operating system making use of a command line/driven interface is: MSDOS.

Graphical user interface (GUI)

Graphical user interfaces (GUIs) are very popular because they are easy to use. Instead of typing in commands, you enter them by pointing at and clicking objects on the screen. Microsoft Windows and Macintosh operating systems use graphical user interfaces. The main features of a GUI include:

Windows – the screen is divided into areas called windows. Windows are useful if you need to work on several tasks.

Icons – these are small pictures used to represent commands, files or windows. By moving the pointer and clicking, you can carry out a command or open a window. You can also position any icon anywhere on your desktop.

Menus – these allow a user to make selections from a list. Menus can be pop-up or pull-down and this means they do not clutter the desktop while they are not being used.

Pointers – this is the little arrow that appears when using Windows. The pointer changes shape in different applications – it changes to an "I" shape when using word-processing software. A mouse can be used to move the pointer around the screen.

Notice that the first letter of each feature in the above list spells out the term WIMP (i.e., **W**indows, **I**cons, **M**enus, **P**ointers).

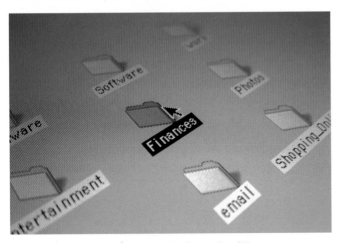

The icons (small pictures) representing folders in a GUI.

Command line interface

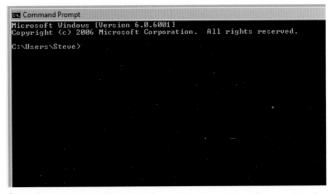

Example of a command line interface where the commands are entered at the prompt.

Post-WIMP interaction

Devices such as phones, PDAs and tablet computers need to be portable and this meant that new interfaces needed to be developed because there was no room for a mouse to be used. These new devices use a user interface called post-WIMP and this uses more than one finger as the input device together with a touch screen. Post-WIMP interaction allows actions such as:

▶▶ Pinching – where you pinch your fingers together to zoom in or spread them further apart to zoom out.

▶▶ Rotating – where you use two fingers – one finger moves up and the other finger moves down to rotate the object such as an image.

▶▶ Swiping – where you swipe your finger over the touchscreen to turn over a page of a document.

Different types of computer

Computers range from the very large to the very small and which one is chosen depends on the type of job it has to perform. There are many different types of computer and the main ones are covered here.

Mainframe computers

Mainframe computers are very sophisticated and powerful computers that are used in very large companies and organizations. They are used to control huge networks of computers often located globally (i.e., in different countries around the world). Airlines, large chains of supermarkets, car manufacturers, government departments, etc., all use mainframe computers.

Mainframe computers have parallel processors, which means that they contain many processors (i.e., the brains of the computer), and can process work simultaneously. This basically means they can process many jobs at the same time. This makes mainframe computers very fast and powerful and capable of running huge networks which may span across many different countries.

◉ KEY WORD

Processor often called the CPU and is the brain of the computer consisting of millions of tiny circuits on a silicon chip. It processes the input data to produce information.

Advantages of mainframe computers:

▸▸ Can process huge amounts of data from scientific experiments which would take too long using a different type of computer.
▸▸ Many processors mean data can be processed very quickly.
▸▸ Can run huge networks with hundreds of terminals by making use of time sharing with each terminal being given a proportion of the mainframe computer's processing time (called a time slice).

Disadvantages of mainframe computers:

▸▸ Need to be placed in a special room with air-conditioning to reduce the large amount of heat produced by the processors.
▸▸ Very expensive to purchase and run.

Uses of mainframe computers:

▸▸ For processing the results of huge amounts of data from large global scientific experiments such as experiments on global warming.
▸▸ For processing transactions (i.e. bits of business) for large companies such as banks, airlines and large government departments.
▸▸ For processing the huge volume of data needed to produce weather forecasts, earthquake and tsunami warnings, etc.

Mainframe computers are powerful computers capable of running hundreds and even thousands of networked computers.

Personal computers (PCs) or desktop computers

A personal computer (or PC as they are sometimes called) is the type of computer that you are most likely to encounter at home or at school.

In many cases personal computers are connected, either with wires/cables or wirelessly, to form networks.

Personal computers can be desktops (i.e., full-sized computers) designed to be used in one place, or laptops designed to be portable and carried and used in different places.

Desktop computers have full-sized keyboards and are designed to be used in one place.

Advantages of desktops compared to laptops:

▸▸ They usually have a better specification (e.g. faster processor, more RAM and ROM, a higher capacity hard disk drive).
▸▸ The keyboard is separate to the screen and both can be adjusted, so the user is less likely to suffer from backache or repetitive strain injury (RSI).
▸▸ They are cheaper to buy and repair.
▸▸ The full-sized keyboard and mouse are easier to use compared with the smaller keyboard and pointing device used with a laptop.

Disadvantages of desktops compared to laptops:

▸▸ They are not portable because they are large and consist of separate components.
▸▸ There are lots of wires which need to be disconnected when moving the computer and this takes time.
▸▸ If work needs to be done at home then you cannot easily take the computer home so you have to spend time copying your files onto removable media.

Laptops

Laptop computers are designed to be portable and used while on the move. They usually use a touch pad instead of a mouse to move the cursor and make selections. Laptops are often used in public places so there is a greater likelihood of them being stolen.

Laptops make use of LCD (liquid crystal displays) which use less power and are light. This is important because laptops use rechargeable batteries when used away from a power supply.

Advantages of laptops compared with desktops:

▸▸ Much smaller and lighter so easily transportable.
▸▸ They can be used with multimedia systems such as data projectors.
▸▸ Can be used on your lap when there is no flat surface on which to work.
▸▸ They have a battery which enables them to be used when away from the mains power supply.

▸▸ Files do not need to be transferred between work and home which saves time.

▸▸ They are Wi-Fi enabled which means they can access networks including the Internet when a signal is available.

Disadvantages of laptops compared with desktops:

▸▸ The smaller keyboards are more difficult to use.

▸▸ The pointing device is harder to use than a mouse, although a mouse can be connected to a laptop and can be used where there is a flat surface.

▸▸ The base of a laptop becomes hot, making it uncomfortable when used on your lap.

▸▸ As they are portable and are used in public places such as coffee shops, airports, etc., they are easily stolen.

▸▸ They have a limited battery life and you need to carry a mains adapter so that they can be recharged.

Palmtops

Palmtop.

A handheld computer, which is smaller than a laptop and can be held in the palm of one hand, is referred to as a palmtop computer. Palmtop computers do not usually have a real keyboard, so selections are made using a pen-like device called a stylus, which is used to touch the touch-sensitive screen. You can have a virtual keyboard which appears on the screen and you touch the letters on the screen to form words.

Personal digital assistant (PDA)

Personal digital assistants (PDAs) are hand-held computers that enable the user to:

▸▸ keep track of meetings, appointments, birthdays, etc.

▸▸ store details of names, addresses, telephone numbers, email addresses, etc.

▸▸ synchronize details with those stored on a desktop or laptop computer

▸▸ browse the Internet

▸▸ send and receive email.

This Blackberry is a PDA/mobile phone that uses a real keyboard rather than a touch screen.

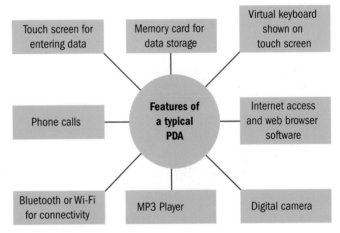

There are a huge number of features of PDAs/palmtops and the distinction between them and mobile phones is almost non-existent.

Advantages of PDAs/Palmtops compared with other types of computer:

▸▸ PDAs are smaller/lighter and are more portable.

▸▸ PDAs are easier to use whilst standing.

▸▸ Many can be used as a mobile phone which means that two separate devices are not needed.

Disadvantages of PDAs/Palmtops compared with other types of computer:

▸▸ PDAs usually have smaller memory which limits the things they can do.

▸▸ PDAs are small and this can make it difficult for some users to use/have small keyboard which can be hard to use/have a small screen which can make it difficult to read text.

▸▸ PDAs are small and so are often lost.

Netbook computers

Netbook computers, or netbooks for short, are smaller, lighter, and less expensive compared to laptop computers.

Advantages of netbooks compared with laptops:

▸▸ Much lighter than laptops (smaller screen and no CD/DVD drive reduces the weight).
▸▸ Longer battery life (owing to the use of less powerful devices such as low-power chips).
▸▸ Cheaper – because some of the more expensive components are left out.

Disadvantages of netbooks compared with laptops:

▸▸ Smaller keyboard can make them more difficult to use.
▸▸ Smaller screen can make the text hard to read.
▸▸ Smaller hard disk so less data can be stored.
▸▸ Hard to upgrade.
▸▸ Lower multimedia quality (e.g. screen, speakers, sound card, etc.).
▸▸ Low performance when doing certain tasks.

Netbooks are smaller and lighter than laptops.

Multifunction devices

Multifunction devices bring together the functions of several devices. For example, by combining PDA and mobile phone technology you can have phones capable of browsing the Internet. There are multifunction devices that offer Internet browsing, satellite navigation, music playing, etc., all using a single device.

Recent developments in ICT

Cloud computing

Cloud computing is Internet-based computing where programs and data are stored on the Internet rather than on the user's own computer. All the resources a user needs to carry out an ICT task are supplied from the Internet.

This service is available now and you can see how it works by accessing the following web site: www.spotify.com.

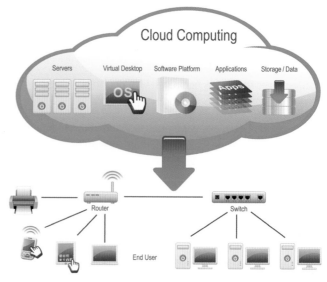

The word "cloud" in "cloud computing" comes from the drawing of a cloud that represents the Internet in many network diagrams.

Advantages of cloud computing:

▸▸ You have instant access to a huge amount of applications software.
▸▸ You do not need to spend large amounts of money for software you only use occasionally.
▸▸ You only need a netbook computer to access resources and these are generally cheaper and more portable than laptop or desktop computers.
▸▸ No need to spend time installing software on your own computer.
▸▸ Your data is held safely on the cloud so there is no need for you to make backups of your own data as these are done automatically for you.
▸▸ You can access data from other devices (e.g. 3G phones)

Disadvantages of cloud computing:

▸▸ You are entrusting your data to a third party for storing and taking backups.
▸▸ The system may be a target for terrorists as the loss of all the data would cause chaos.
▸▸ There is a danger of abuse by hackers and people who introduce viruses onto the system.
▸▸ Use of the system could be expensive as resources are supplied after you pay a subscription.
▸▸ Loss of privacy as a company will now store all your private materials such as letters, accounts, emails, etc.

Applications of cloud computing

Streaming music over the Internet

Music streaming allows you to store your music remotely and then listen to it using the Internet. This would save you having to download music onto your computer or MP3 player. With the new system you can log onto a web site using wireless Internet (Wi-Fi) and then you could listen to tracks using Internet streaming. This means that the music will be stored "in the cloud" on the Internet and you would listen to tracks and pay a monthly subscription.

Google docs

Google docs is an application of cloud computing that provides:

▸▸ free web-based software such as word-processor, spreadsheet, slide shows, etc.
▸▸ data storage that allows documents to be accessed from any device that can access the Internet.

Google docs is ideal for people who are working together on a document. They can each access the document and add their material or make changes to it.

E-books and newspapers

E-books and newspapers are starting to replace traditional paper books and newspapers. E-books and e-newspapers can be read using a special reader, called an e-reader, or by a computer such as a netbook or i-Pad. The book or newspaper is downloaded using the Internet (usually wirelessly) to the reader or computer. Usually, you pay a subscription for the service.

The main advantages of e-books and newspapers are:

▸▸ Storage of thousands of books on one device (e.g. Kindle reader).
▸▸ They are more environmentally friendly than paper books/newspapers.
▸▸ You have much more choice.
▸▸ You can download books/newspapers almost instantly.

A businessman reading a newspaper using a touchpad computer.

Important note Windows is an interface which allows easy user interaction with a multitasking graphical user interface (GUI) operating system as well as other software. MS Windows is a branded operating system owned by the company Microsoft.

QUESTIONS B

1 Graphical user interfaces are very popular interfaces that are used with computers, mobile phones, and other portable devices.

   a Give **three** features of a graphical user interface.

      (3 marks)

   b Other than a graphical user interface (GUI) give the name of **one** other type of user interface. *(1 mark)*

2 All computers need an operating system.

   a Explain what an operating system is. *(2 marks)*

   b List **three** different functions of an operating system.

      (3 marks)

REVISION QUESTIONS

1 Give **two** differences between a desktop computer and a laptop computer. **[2]**

2 Desktop computers consist of a number of components. Explain the purpose of each of the following components:

   a Central processing unit (CPU) **[1]**

   b Internal memory devices **[1]**

   c Backing storage **[1]**

3 Give the names of:

   a **Two** input devices. **[2]**

   b **Two** output devices. **[2]**

   c **Two** backing storage devices. **[2]**

4 Computers come in all sizes and can be used in different situations.

   a Describe what is meant by a desktop computer and explain how it differs from a laptop computer. **[3]**

   b Netbook computers are becoming very popular. Give **two** ways in which a netbook computer differs from a laptop computer. **[2]**

5 A salesperson travels to newsagents taking orders for sweets using a PDA.

   a Give the meaning of the abbreviation PDA. **[1]**

   b Give **one** advantage of using a PDA rather than a laptop computer. **[1]**

6 Mainframe computers are used by large organizations.

   a Give **two** features of a mainframe computer. **[2]**

   b Give **two** uses for mainframe computers. **[2]**

7 The latest portable devices are small and many use touchscreens rather than a mouse. They are said to use post-WIMP interaction. Explain how each of the following post-WIMP interactions is used.

   a Rotating. **[1]**

   b Pinching. **[1]**

Test yourself

The following notes summarize this topic. The notes are incomplete because they have words missing. Using the words in the list below, copy out and complete the sentences A to M underlining the words that you have inserted. Each word may be used more than once.

software   ROM   information   hardware   desktop
output   CPU   input   RAM   pinching   applications

A Computer systems consist of two main parts: hardware and _____.

B If you can physically touch it then it is _____.

C The _____ is a piece of hardware that is the brains of the computer and it turns data into _____.

D Computer hardware is useless without the _____ which is used to give it instructions as to what to do.

E There are two types of software called operating system software and _____ software.

F A computer which has a full-sized keyboard and full-sized screen and is normally used in one place is called a _____ computer.

G _____ devices such as the keyboard and mouse are used to enter data into the computer for processing.

H After data has been processed, the results of processing are passed to an _____ device.

I _____ is fast permanent memory used for holding instructions needed to start the computer up.

J _____ is a fast temporary memory where programs and data are stored only when the power is supplied.

K _____ is held on a computer chip and is called non-volatile memory because it does not lose its contents when the power is turned off.

L _____ is held on a computer chip and is called volatile memory because the contents disappear when the power is turned off.

M Post-WIMP interaction is a user interface that allows _____ and rotating.

EXAM AND EXAM-STYLE QUESTIONS

1 Ring **two** items which are storage media   *[2]*

Magnetic tape   Touch pad   Flash memory card

OCR   Chip reader   Light pen

2 There are two types of internal memory called ROM and RAM. Describe the differences between ROM and RAM.

[4]

3 Ring **two** items which are examples of software.   *[2]*

Mouse   DVD   Web browser

Operating system   Laser printer   CD ROM

4 Complete each sentence using **one** item from the list.   *[5]*

Router   Output   Communication   Software

Processor   Microphone   ROM   RAM   Input

a .... means the programs that supply the instructions to the hardware to tell it what to do.

b .... devices such as keyboards, mice and scanners are used to supply data to the computer.

c .... devices are hardware such as printers, speakers and screens.

d .... is memory which is used to hold the boot program needed to start the computer up when first switched on.

e .... is memory held on a chip that can have its contents changed by the user.

5 Give **three** functions of an operating system.   *[3]*

6 Graphical user interfaces are very popular particularly with PDAs and multifunction devices.

a Give **two** advantages of a PDA compared to a laptop computer.   *[2]*

b Give **two** disadvantages of a PDA compared to a laptop computer.   *[2]*

c Give **three** features of a graphical user interface.   *[3]*

Input and output devices

Input devices such as a keyboard and mouse/touchpad are available with nearly all computers. There are, however, other input devices which reduce the amount of work and improve accuracy when entering data into the computer for processing.

Once data has been processed, the results of the processing, called the output, will need to be produced. Various output devices such as printers, screens, speakers, etc., are used for output.

In this chapter you will be covering both input and output devices.

The key concepts covered in this chapter are:
▸▸ Identification of the key input devices, their uses, and advantages and disadvantages
▸▸ Identification of the key output devices, their uses, and advantages and disadvantages

Input devices

Input devices are those hardware devices that are used for the entry of either instructions or data into the computer for processing. There are many different input devices and which one is chosen depends mainly on the task being performed. For the examination you will need to know how each of the following input devices is used and their relative advantages and disadvantages.

Keyboard

Keyboards are used to enter text into word-processed documents, numbers into spreadsheets, text into online forms, and so on. All computers come with keyboards so they are the natural choice for the entry of text that cannot be copied from other sources.

The main advantages of using keyboard entry are:

▸▸ Text appears on the screen as you type so can easily be checked for accuracy.
▸▸ Most computers come with them so there is no extra hardware to buy.
▸▸ They are ideal for applications such as word-processing or composing emails where you have to create original text.
▸▸ Can be used for other instructions (e.g. Ctrl+P to print).

The main disadvantages of using keyboard entry are:

▸▸ It is a slow method for entering large amounts of text compared to methods such as dictating or scanning text in.
▸▸ It can be inaccurate as it is very easy to make typing mistakes.
▸▸ It is a frustrating method to use if you do not have good typing skills.
▸▸ Typing for a long time can cause a painful condition called repetitive strain injury (RSI).

Numeric keypad

Numeric keypads are used where only numeric data is to be entered. For example, you may see them being used in ATM

This separate numeric keypad is used in a bank to enter amounts when a customer pays into their account.

machines (Automatic Teller Machines) (i.e., cash dispensers) to enter the PIN (Personal Identification Number) and the amount that needs to be withdrawn.

They are also used in chip and PIN machines where you have to insert a credit/debit card and then enter your PIN to pay for goods or services.

The main advantage of using a numeric keypad is:

▸▸ It is smaller, making it easier to carry or move, e.g. at a shop counter.

The main disadvantages of using a numeric keypad are:

▸▸ Small keys might be hard to see.
▸▸ Keys might be arranged differently, e.g. on a phone.

Pointing devices

Pointing devices are needed to make selections from a graphical user interface. In order to move the pointer onto a button/icon/hyperlink, etc., it is necessary to use one of the following devices:

▸ *Mouse* – needs no introduction. The pointer moves in response to the movement of the mouse. Buttons are used to make selections. The left mouse button is used to make selections whilst the right button is used to display a drop-down menu. A scroll button/wheel is used to help move quickly through long documents.

Advantages in using a mouse include:
▸ Enables rapid navigation through an application, e.g. a web browser.
▸ Small, so uses a small area on the desktop.

Disadvantages in using a mouse include:
▸ Hard to use if the user has limited wrist/finger movement.
▸ Needs a flat surface so can't be used "on the move".

▸ *Touchpad* – these are used on laptop computers when there is no flat surface to use a mouse. Many people find them more difficult to use and prefer to have a mouse handy to attach when there is a flat surface. They also have buttons to act like mouse buttons.

Touch pads are used as an alternative to a mouse when there is not much space for a mouse to be used.

Advantages in using a touchpad include:
▸ Can be used "on the move" where there is no flat surface on which to move a mouse.
▸ There is no extra device to carry as the touchpad is built into the computer.
▸ It is faster to navigate through an application compared to using a keyboard.

Disadvantages in using a touchpad include:
▸ They are hard to use if the person using the touchpad has limited wrist/finger movement.
▸ Controlling the pointer is harder compared to using a mouse.
▸ They are hard to use for actions such as "drag and drop".
▸ They tend to be used in more cramped conditions so their long use could lead to health problems.

▸ *Trackerball* – these are a bit like an upside-down mouse and you move the pointer on the screen by rolling your hand over the ball. They are used by people with poor motor skills such as the very young or people with disabilities.

A trackerball is an alternative to a mouse.

Advantages in using a trackerball include:
▸ They are easier to use than a mouse if the user has wrist or finger problems.
▸ They allow faster navigation compared to a mouse.
▸ They are stationary as it is the ball that moves.

Disadvantages in using a trackerball include:
▸ The cost, as trackerballs are not normally included with a computer system.
▸ Time is needed to get used to using them.

Remote control

Remote controls are input devices because they issue instructions to control devices such as TVs, video players/recorders, DVD players/recorders, satellite receivers, Hi-Fi music systems, data projectors or multimedia projectors.

Remote controls are used to issue control instructions to devices.

Advantages in using a remote control include:
▸ The user can operate a device wirelessly from a short line of sight distance where a device's controls are difficult to reach (e.g., a projector positioned on a ceiling).

▸ Can control devices safely from a distance (e.g. a robot used to investigate a bomb).

▸ Enables people with disabilities to operate devices at a distance.

Disadvantages in using a remote control include:

▸ Remote control usually uses infra-red signals and these signals can be blocked by objects in their path so the remote control may not operate.

▸ Remote controls need batteries to operate and need constant replacement.

▸ Small buttons can be pressed by mistake and this can alter the settings of the device and make it hard for the user to go back to the original state.

Joystick

Joysticks are used mainly for playing computer games. They can also be used by a pilot to fly an aeroplane or with a mock-up of plane in a flight simulator. They can also be used in car driving simulators. The stick moves on-screen in the same way as a mouse, and buttons are used to select items.

This games controller uses two joysticks.

Advantages of joysticks:

▸ They are ideal for quick movement.

▸ They can be used by disabled people because they can be operated by foot, mouth, etc.

Disadvantages of joysticks:

▸ Entering text is very slow as you have to select individual letters (e.g. the name of a high scorer in a game).

▸ They are not much use for entering large amounts of text.

▸ You have to purchase them separately from the computer system.

Touch screen

Touch screens are displays that can detect the presence and location of a touch to the screen. Usually a finger is used to make selections on the screen, although some systems allow several fingers to be used, for example re-sizing an image on the screen by moving two fingers closer together or further part. Touch screens can be found in information kiosks, all-in-one computers, tablet computers, PDAs and smartphones.

Touch screens are popular in restaurants and shops for inputting orders/sale details because employees need little training to use them. They are also features of some point of sale terminals (i.e., computerized tills). They are ideal for tourist information kiosks, transport information, airport self-check in or ticket collection points.

Many mobile phones have touch screen interfaces.

Sat navs use touch screens.

Computerized ticket dispensers at train stations make use of touch screens.

Advantages of touch screens:

- They are very simple to use so can be used by almost anyone.
- It is easy to choose a particular option.
- They are easier to use whilst standing compared to using a keyboard.
- They are ideal where space is limited such as on a smart phone, PDA, etc.
- They are tamper-proof so other data cannot be entered which could corrupt the system.

Disadvantages of touch screens:

- The screen can get quite dirty and this can make items on the screen hard to see.
- There is a danger that germs can be spread with everyone touching the same screen.
- On small screens the icons can be hard to see or select.
- They often cost more than alternative input devices such as a keyboard.

Magnetic stripe readers

Magnetic stripe readers read data stored in the magnetic stripes on plastic cards such as loyalty cards. The stripe contains data such as account numbers and expiry date.

They are used for reading data off credit/debit cards where a chip and PIN reader is not available. They are also used in ID cards where the card is swiped through a reader to gain access to buildings and rooms. Other uses include pre-payment cards for using services such as the Internet, photocopiers, etc.

Card with magnetic stripe.

Advantages of magnetic stripe readers:

- Faster input of data by swiping which is faster than typing in the data.
- Can be used as an alternative method to chip and PIN for credit/debit cards.

- Avoids possible typing errors which could be introduced by keying in.
- Stripes are not affected by water so they are robust.

Disadvantages of magnetic stripe readers:

- Magnetic stripes can only store a small amount of data.
- Magnetic stripes can be damaged easily by magnetic fields.
- Magnetic stripes can be duplicated relatively easily (called card cloning) and this leads to card fraud.
- The cards cannot be read at a distance as you have to put them into the reader and swipe them.

Chip readers and PIN pads

Chip readers are the devices into which you place a credit/debit card to read the data which is encrypted in the chip on the card.

The chip on a credit card.

The PIN pad is the small numeric keypad where the personal identification number (PIN) is entered and the holder of the card can be verified as the true owner of the card.

The main use of chip and PIN is to read card details when making purchases for goods or services where the cardholder is present to input the PIN.

The chip reader and the PIN pad commonly referred to together as chip and PIN.

Advantages of chip readers and PIN pads:

▸▸ They have reduced fraud as the true cardholder has to input their PIN.
▸▸ Chips are harder to copy compared to a magnetic stripe.
▸▸ The storage capacity for data on a chip is much higher than that for a magnetic stripe.
▸▸ No problem with wear as on a magnetic stripe.

Disadvantages of chip readers and PIN pads:

▸▸ Not all countries can used chip and PIN, so they have to use magnetic stripe readers.
▸▸ People can forget their PIN.
▸▸ People can look when the PIN is being entered (called "shoulder surfing").

Chips are also found in ID cards, some passports, public transport tickets and these are fed into readers to allow access, display information stored on the chip, etc.

Scanners

Scanners can be used to scan in photographs and other images on paper to put into documents or web pages. Scanners therefore provide a way of digitizing hard copy documents. They can also be used for scanning text into a word-processing or other package thus saving having to re-type the text. Scanning and recognizing text in this way is called OCR (optical character recognition).

Advantages of scanners:

▸▸ They are the only way of digitizing old paper-based images.
▸▸ They can be used in conjunction with special optical character recognition software to scan and enter text into word-processing or other software without the need for re-typing. This is called OCR (optical character recognition).
▸▸ They can be used to digitize old documents for archive purposes.
▸▸ They can be used with software to repair/improve old photos.

Disadvantages of scanners:

▸▸ They take up a lot of space.
▸▸ It can take a long time to digitize all the pages in a long document.
▸▸ The quality of the scanned item might be poor depending on scanner resolution.

Scanners can be used to digitize old documents such as maps, old pictures, and photographs.

"I HAVE TO STAY HOME TONIGHT AND HELP MY DAD WITH HIS NEW CAMERA PHONE. WE NEED TO DELETE 750 PICTURES OF HIS HAND."

Digital cameras

Digital cameras store the digital photographs on a memory card and display the picture on a screen. The digital image can then be saved on a computer or edited using photo-editing software and then saved. Digital cameras can often be connected directly to a photo printer (a printer designed specifically for printing photos) to print out the photographs without the need for a computer. Nearly all mobile phones also have a digital camera facility.

The dots making up a digital image are called pixels and generally the higher the number of pixels the more detailed the photograph. This also means the file size will be larger and fill more space on the memory card. Digital video cameras are available which take moving images and sound and are very useful for adding to web sites, presentations, and other multimedia materials.

Advantages of digital cameras:

▸▸ No film to develop so quicker to produce the photo.
▸▸ No need to use chemicals in developing films.
▸▸ A large number of photographs can be taken of the same thing and the best one can be chosen.
▸▸ The images are in a form that can be placed in documents, in presentations and on web sites, etc.
▸▸ The digital images they produce are easily sent from place to place using phones, email attachments, etc.
▸▸ They can be placed on photo or video sharing sites for others to see.
▸▸ Can use software to improve photo (e.g. remove red-eye).

Disadvantage of digital cameras:

▸▸ The photographs may not be as high quality as those produced using film. You are limited by the number of pixels (i.e. small dots) that make up the image. The number of pixels used to make up the image is called the resolution.
▸▸ A photographer needs to have computer skills to use digital photographs properly.

Microphones

Microphones are input devices that allow sound signals to be converted into data. They are used to digitize (i.e. convert the sound wave into a series of pulses consisting of 0's and 1's) sound so that it can be stored and processed by the computer. Microphones are used to give instructions to the computer or enter data using special software called voice recognition software.

Here is what a microphone allows you to do:

- You can instruct without using hands (e.g. hands-free phone in a car).
- You can dictate letters and other documents directly into your word-processor or email package. This is called voice recognition.
- You can take part in videoconferencing and can issue instructions verbally to your computer instead of typing them in.
- You can add narration (i.e., spoken words) to web sites, presentations, and other multimedia products.
- You can input speech for VoIP (Voice over Internet Protocol which is a cheap way of conducting phone calls using the Internet) and videoconferencing systems.

Advantages of microphones:

- Can be used by disabled to input data/instruct.
- The only way of inputting speech that cannot be copied from elsewhere.
- Microphones are very cheap to buy.
- Can be faster to speak instructions rather than type them.

Disadvantage of microphones:

- Background noises can cause problems with voice recognition systems.
- Voice recognition is not completely accurate so mistakes may occur.
- Sound files, when stored, take up a lot of disk space.

Sensors

Sensors are able to measure quantities such as temperature, pressure and amount of light. The signals picked up by the sensors can be sent to and then analysed by the computer. Sensors can:

- be connected to the computer directly
- be found in lots of devices such as burglar alarms, central heating systems, washing machines, etc., – as well as sensing, they are also used to control the device in some way
- be used to input data for the control of certain devices.

Advantages of using sensors:

- The readings are more accurate than those taken by humans.
- The readings can be taken more frequently than by a human.
- They work when a human is not present so cost less.
- They can work in dangerous environments (e.g. mines or nuclear reactors).

Disadvantages of using sensors:

- Cost to purchase.
- Dirt and grease may affect performance.

Temperature sensors – used to monitor temperature. When the temperature data is sent to the computer/microprocessor it can be used to control heaters, coolers, windows, etc. Temperature sensors can be used to control automatic washing machines, central heating systems, air conditioning systems, greenhouses, and environmental monitoring systems.

Pressure sensors – pressure sensors measure liquid pressure (the pressure in a liquid increases with depth) and pressure when something is pressing down on a pad. Some pressure sensors measure atmospheric pressure which is an important quantity for predicting the weather. Pressure sensors can be used in burglar alarm systems to detect the pressure exerted by a burglar. They are also used in washing machines to detect when the water has reached a certain depth.

Other uses for pressure sensors include robotics (so that robot arms can pick things up without squashing them), production line control, and environmental monitoring (e.g., flood warnings in rivers).

Light sensors – these are used to measure light intensity and can be used in security lights which come on in the dark and go off when it gets light. Light sensors can be used for burglar alarm systems, production line control, scientific experiments, and environmental monitoring (e.g. the conditions in a greenhouse).

Humidity/moisture sensors – these are used to measure the amount of moisture in the air or soil. They are used in greenhouses to ensure ideal growing conditions for plants.

Graphics tablet

A graphics tablet is a flat board (or tablet) which you use to draw or write on using a pen-like device called a stylus. Graphics tablets are ideal for inputting freehand drawings. They can also be used for retouching digital photographs.

Graphics tablets are used for specialist applications such as computer-aided design (CAD) (e.g. kitchen design) and have a range of specialist buttons for certain shapes and items. You can add these items by clicking on them with the stylus.

A graphics tablet.

Advantages of graphics tablets:

▶▶ Used in countries such as Japan and China where graphical characters are used instead of letters for words.
▶▶ It is more accurate to draw freehand on a tablet rather than use the mouse to draw.
▶▶ The icons/buttons are on the graphics tablet rather than the screen leaving more space on the screen for the design/drawing.

Disadvantage of graphics tablets:

▶▶ Specialist tablets are expensive.

Optical mark reader (OMR)

Optical mark readers use paper-based forms or cards with marks on them which are read automatically by the device. OMR readers can read marked sheets at typical speeds of 3000 sheets per hour. OMR is an ideal method for marking multiple choice question answer sheets for examinations and tests. The students mark the bubbles or squares by shading them in and the reader can read and process the results at high speed.

OMR is used to assess customer satisfaction with a product or service.

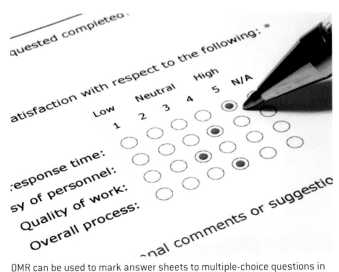

OMR can be used to mark answer sheets to multiple-choice questions in exams.

Advantages of optical mark recognition include:

▶▶ Only need one computer and optical mark reader to read the marked sheets.
▶▶ If the forms have been filled in correctly, then almost 100% accuracy can be achieved as there are no typing errors.
▶▶ The computer is fast at reading the forms and analysing the results. This is particularly important if a large volume of data needs to be input and processed in a short space of time (e.g. tests and assessments).
▶▶ Less expensive then employing workers to type the answers in.

Disadvantages of optical mark recognition include:

▶▶ Only suitable for capturing certain data – data needs to be in a form where there are tick-style answers.
▶▶ Reject rate can be high – if you have not given precise instructions, users may fill in the forms incorrectly, which will lead to high rejection rates.
▶▶ If the form used is creased or folded it may be rejected or jam the machine.

Suitable applications for optical mark recognition include:

▶▶ voting forms
▶▶ lottery tickets
▶▶ tests/assessments
▶▶ school/college attendance registers.

Optical character reader (OCR)

An optical character reader works by scanning an image of the text and then using special recognition software to recognize each individual character. Once this is done the text can be used with software such as word-processor, desktop publishing or presentation software.

OCR is used for reading account numbers and details on utility (gas, electricity, water or telephone) bills. It is also used where there is a large amount of text that needs inputting such as where the text of a book needs to be digitized or for forms used in passport applications.

OCR is used in conjunction with CCTV cameras to recognize car registration plates so that the vehicle can be checked to see if it is taxed and has valid insurance.

Advantages of OCR:

▶▶ A fast way of inputting text that has been printed on paper.
▶▶ Avoids having to type the text in, which reduces the risk of RSI (repetitive strain injury).
▶▶ Can recognize handwriting so can be used to handwrite notes on a tablet computer and convert them to word-processed text.
▶▶ Can avoid typing errors.

Disadvantages of OCR:

▶▶ Text needs to be clearly typed or written (e.g. handwriting is poorly read).
▶▶ The forms may be rejected if they are incorrectly filled in.

Bar code reader

Bar codes are a series of light and dark bars of differing widths. They are used to input the number using a bar code reader, which appears below the barcode which is then used to look up the item details in a database.

The bar code is used to input the number below the bar code without having to type it in.

Using the code the system can determine from a product database:

- ▶ the country of origin
- ▶ the manufacturer
- ▶ the name of the product
- ▶ the price
- ▶ other information about the product.

Suitable applications for bar code recognition include:

- ▶ producing itemized customer receipts in supermarkets/stores
- ▶ warehouse stock control systems
- ▶ tracking the progress of parcels during delivery
- ▶ recording books loaned to members in a library
- ▶ luggage labelling at airports.

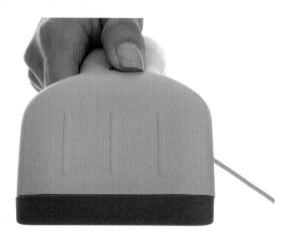

A hand-held scanner being used to input the numbers stored in the bar code.

Advantages of bar code input:

- ▶ Faster than typing in the number.
- ▶ Greater accuracy – compared to typing in long codes manually.
- ▶ Can be read from a distance – useful in wholesalers where the goods are often too heavy to be removed from the trolley.
- ▶ Can record barcode and time/date at the same time, which is useful for tracking parcels or bags at an airport.
- ▶ Prices can be changed by altering the database so you do not have to have price stickers on each item.

Disadvantages of bar code input:

- ▶ Bar codes can sometimes be damaged – this means having to type in the long number underneath manually.
- ▶ Expensive – the laser scanners in supermarkets are expensive, although hand-held scanners are relatively cheap.
- ▶ Only a limited amount of data can be stored in a bar code – for example sell by dates cannot be stored.
- ▶ Must have computers and network infrastructure at each point where the bar code may be read, which adds to the expense.

Video camera

Most modern video cameras are digital, which means the files produced are digital and can therefore be transferred to a computer and saved or edited and then saved. Like any digital file, the file may be added to presentations, web sites, etc. You can also add the video to sites such as YouTube so that you can share the video with others.

Advantages of video cameras:

- ▶ Can capture both still and moving images.
- ▶ Easy to transfer video to the computer.
- ▶ Can store many images/videos until later for editing.

Disadvantages of video cameras:

- ▶ Can erode privacy as CCTV cameras are present in many places.
- ▶ It is very difficult to produce good video without training.
- ▶ Editing is likely in order to make a good video and this can be complicated.
- ▶ Can be expensive to buy.
- ▶ The picture quality of the video is determined by the number of pixels used (i.e. the resolution) and this may be low.

Web cam

Web cams are digital cameras that can take both still and video images which can then be transferred to other computers or saved. They are used for conducting conversations over the Internet where you can see the person you are talking to.

Web cams are now included with many computers and are easily added to computers not having one. The uses for web cams are many and include:

- To conduct simple videoconferencing.
- Allow parents to see their children in nurseries when they are at work.
- To record video for the inclusion on web sites.
- Allow people to view the traffic in local road systems.
- Allow people to view the actual weather in a place they are going on their holidays.

Advantages of web cams:

- Can see the reactions or people as you are talking to them.
- Parents can see their children and grandchildren and speak to them if they do not live near.

Disadvantages of web cams:

- Limited extra features to improve image quality.
- The picture quality can be poor at low resolution.
- Generally have a fixed position so do not move around

Light pen

Light pens are ideal where space is limited and are used as an alternative input device to a mouse or graphics tablet. Light pens are used for producing freehand drawings directly on a screen or editing existing drawings. They can also be used to point to and select items on the screen in a similar way to a touch screen but with greater accuracy.

Advantages of light pens:

- Enable users to draw directly on the screen.
- Useful where there is no space to use a mouse or graphics tablet.

Disadvantages of light pens:

- Using a light pen can be uncomfortable for the user as they have to keep their hand held in a similar position for long periods whilst drawing.
- Can only be used with a CRT (cathode ray tube) screen and not an LCD (liquid crystal display) or TFT screen.

Web cams are good fun because you can see the person you are talking to.

⊙ KEY WORDS

Hard copy documents that are on paper.
Sensors devices which measure physical quantities such as temperature, pressure, etc.
Touch screen a special type of screen that is sensitive to touch. A selection is made from a menu on the screen by touching part of it.
Voice recognition the ability of a computer to "understand" spoken words by comparing them with stored data.

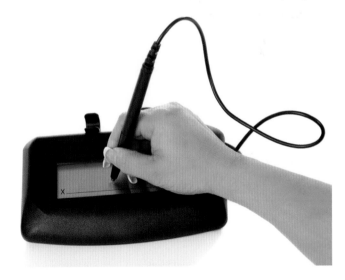

A light pen being used.

1 **a** Tick **six** boxes to show which of the following are input devices. *(6 marks)*

Tick 6 boxes only

Graphics tablet	☐
Colour laser printer	☐
Mouse	☐
Microphone	☐
Speakers	☐
LCD screen	☐
Touch screen	☐
Digital still camera	☐
Magnetic hard disk drive	☐
CD ROM drive	☐
Web cam	☐

b Name **two** output devices given in the table above. *(2 marks)*

2 **a** Explain the purpose of an input device. *(1 mark)*

b Give the names of **two** input devices that would be used by a desktop computer. *(2 marks)*

c Give the name of **one** input device that would be found on a laptop computer that you would not find on a desktop computer. *(1 mark)*

3 Here is a list of input devices:
Optical mark reader
Magnetic stripe reader
Keyboard
Scanner
Microphone

Write down the name of the input device that is most suited for each of these tasks:

a Inputting a short email message. *(1 mark)*

b Reading the numbers on a large number of lottery tickets at high speed. *(1 mark)*

c For inputting an old photograph. *(1 mark)*

d For inputting loyalty card details when a customer makes purchases in a supermarket. *(1 mark)*

e For dictating a novel into a word-processing package. *(1 mark)*

f For reading lots of multiple-choice answer sheets. *(1 mark)*

4 Laptop computers enable people to do work or keep in touch when they are travelling.

Give the name of **two** input devices normally used with a laptop computer. *(2 marks)*

5 The image shows an input device being used.

a Give the name of this input device. *(1 mark)*

b Explain how this device is used. *(3 marks)*

c Give **two** applications where this device could be used. *(2 marks)*

Output devices

After data entered into the computer has been processed, the resulting information needs to be output. There are many output devices, with a screen (monitor) and a printer being the most popular.

Monitors (screens)

Monitors come in lots of different sizes and they are useful for displaying information that is not needed as hard copy (i.e., printouts on paper).

CRT monitors

CRT stands for cathode ray tube and these monitors are the older fatter monitors that you still see being used. They take up more space than the more modern thinner TFT screens.

Advantages of CRT monitors:

▸▸ Can be used with a light pen to produce and edit drawings using CAD software.

▸▸ Sometimes they are used where several people need to view the screen at the same time, for example where several designers are viewing a prototype as they have a wide angle of viewing.

Disadvantages of CRT monitors:

▸▸ They are bulky and take up a large amount of space on the desk.

▸▸ They generate a lot of heat and can make rooms hot in the summer.

▸▸ Glare on the screen can be a problem.

▸▸ They are much heavier and present a safety problem when being moved due to their weight.

▸▸ Flicker can cause headaches and eyesight problems.

Many CRT monitors are being replaced by TFT monitors. Like all computer equipment, they should be recycled.

TFT (thin film transistor) monitors

These are the thin flat panel screens you see and they are the more modern type of monitor. They are used with newer desktop computers and laptops. The screen is made up of thousands of tiny pixels. Each pixel has three transistors – red, green and blue. The colour is generated by the intensity of each.

Advantages of TFT monitors:

▸▸ They are very light and this is why they are used in laptop computers.
▸▸ They are thin so they do not take up much desk space.
▸▸ They are cheaper to run as they use less power than CRT monitors.
▸▸ They can be easily positioned to suit the user because they are light.
▸▸ The radiation given off is much less than that for a CRT monitor.
▸▸ They do not get as hot as CRT monitors and energy is not wasted in the summer using air conditioning to cool offices down.
▸▸ They do not create glare and may be less likely to cause eye strain and headaches.
▸▸ They are ideal where only one person needs to view the screen.

Disadvantages of TFT monitors:

▸▸ They are not easy to repair, so if they go wrong it is usually cheaper to replace them, which is not very environmentally friendly.
▸▸ They have a narrow viewing angle so the image is not as clear as CRT when viewed from the side. They are less useful if many people are viewing the screen.

Multimedia projectors

Multimedia projectors are used to project what appears on the screen of a computer onto a much larger screen enabling an audience to view it. They are used for training presentations, teaching, advertising presentations, etc. They can also be used to provide a larger picture from televisions and video/DVD players.

Multimedia projector.

Advantages of multimedia projectors:

▸▸ Many people can view what is being presented.
▸▸ Can show video to a large audience who may be seated some distance from the screen.

Disadvantages of multimedia projectors:

▸▸ The image quality may not be as high as when seen on a computer screen at normal size.
▸▸ The cooling fans on the projector create noise which can distract.
▸▸ The image needs a darkened room otherwise it will appear dim.

Printers

Printers are used to provide users with output in hard copy form. This means the output is printed on paper.

Laser printers

Laser printers are the type of printer mainly used by businesses and organizations mainly because of their high speed. Because of their high speed they are usually chosen as printers used for networks. Most laser printers print in black and white. Although you can buy colour laser printers, they are relatively expensive.

Advantages of laser printers:

▸▸ Very quiet – important when used as a network printer in an office where phones are also being used.
▸▸ Supplies last longer – the toner cartridge lasts longer than inkjet cartridges.

- ▸▸ High printing speed – essential to have high speed if lots of people on a network are using the one printer.
- ▸▸ Very reliable – fewer problems compared to inkjet printers.
- ▸▸ No wet pages that smudge – inkjet pages can smudge but there is no such problem with a laser printer.
- ▸▸ They are cheaper to run (i.e., they have a lower cost per page compared to inkjet printers).
- ▸▸ High quality printouts.

Disadvantages of laser printers:

- ▸▸ More expensive to buy – but they are cheaper to run.
- ▸▸ Colour lasers are very expensive to buy.
- ▸▸ Power consumption is high.
- ▸▸ Size as most laser printers are larger than inkjet printers.

Inkjet printers

Inkjet printers are popular with home users because they are relatively cheap to buy. However, they are more expensive to run, because of the high cost of the ink cartridges. They work by spraying ink onto the paper and can produce very good colour or black and white printouts. They are ideal printers for stand-alone computers but are less suitable for networked computers as they are too slow.

Inkjet cartridge refills can be expensive.

Advantages of inkjet printers:

- ▸▸ Relatively small compared to laser printers.
- ▸▸ They do not produce ozone or organic vapours which can cause health problems.
- ▸▸ High quality print – ideal for printing photographs, brochures, and illustrated text.
- ▸▸ Quietness of operation – this is important in an office as telephone calls or conversations with colleagues can be conducted during printing.
- ▸▸ Inkjet printers are usually cheaper than laser printers.

Disadvantages of inkjet printers:

- ▸▸ High cost of the ink cartridges – this is ok for low volume work but for large volume it is cheaper to use a colour laser printer.
- ▸▸ Ink smudges – when the printouts are removed the paper can get damp which tends to smudge the ink.
- ▸▸ Cartridges do not last long – this means they need to be constantly ordered and replaced frequently.

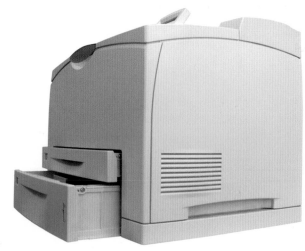

Laser printers are ideal for use in places where lots of copies are needed quickly.

Dot matrix printers

Dot matrix printers are used in offices and factories where multi-part printouts are needed and noise is not an issue. For example, one copy of an invoice could be kept by the sales office and one copy could be sent to the customer and the other given to the factory. They are impact printers and work by hitting little pins against an inked ribbon to form the characters on the paper, which creates a large amount of noise.

Dot matrix printers may seem out-of-date but they are ideal in some situations. You will see them being used to print out invoices in warehouses, car part stores, and garages. You can also see them being used in pharmacies for the printing of labels.

Dot matrix printers use an inked ribbon.

Advantages of dot matrix printers:

- ▸▸ Can be used to print multi-part stationery – this is useful where the printout needs to be sent to different places.
- ▸▸ Can be used with continuous stationery – this makes it ideal for stock lists, invoices, etc.
- ▸▸ Relatively slow.
- ▸▸ Cheaper to run than inkjet printers.

Disadvantages of dot matrix printers:

▸▸ Very noisy – the pins make a lot of noise when they strike the inked ribbon.

▸▸ Low quality printouts. The text can look "dotty".

▸▸ Unsuitable for printing graphics.

▸▸ It is possible to get colour dot-matrix printers but they are not used very often now that other colour printers are cheaper and produce higher quality printouts.

▸▸ More expensive to buy than inkjet printers.

Graph plotter

Graph plotters are ideal for printing designs, plans, and maps. They produce very precise scale drawings. They are ideal for printing out designs that have been produced using CAD (computer-aided design) software. They are also ideal for printing out large printouts, for example on A0-sized paper.

Advantages of graph plotters:

▸▸ Can be used with larger-sized paper or card.

▸▸ Can be adapted and used as a cutter to cut out designs.

▸▸ Produces high quality printouts.

Disadvantages of graph plotters:

▸▸ Slow to complete the image.

▸▸ Expensive to buy and maintain.

Speakers

Any application which requires sound will need speakers or ear phones to output the sound. Applications which use sound include multimedia presentations and web sites. Many online encyclopaedias make use of sound with explanations, famous speeches, music, etc.

People also use their computer for playing their CDs whilst working or maybe listening to the radio using the Internet. Speakers are also needed to output the sound when you watch a DVD film on the computer. With laptop computers the speakers are usually built in. Desktop computers may have basic speakers included but many people choose to upgrade these to better sound quality speakers.

Speakers enable users to use the multimedia features of computer systems.

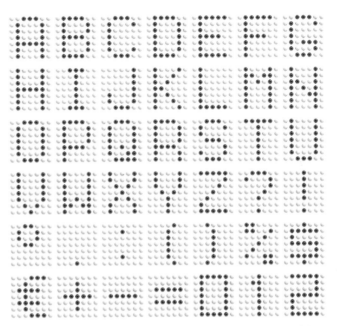

With a dot matrix printer, characters are formed by an arrangement of dots.

Output devices as control devices in control applications

Computers can be used for control in many devices and the computer sends control signals which turn control devices on or off. Control devices can be motors, buzzers, heaters, lamps, etc.

Actuators

Actuators are any device that responds to signals from a computer used in a control situation that produces movement. For example, a switch or valve that needs to be opened or closed is an example of an actuator.

An actuator would be needed to allow a computer to turn a motor or light on or off.

Motors

Computers can issue control signals to turn electric motors on or off. Some special motors, called stepper motors, only turn through a certain angle depending on the signals they receive from the computer.

Here are some applications of motors being used in control systems:

▸▸ Motors used in automatic washing machines to turn valves on/off to allow the water in/out.

▸▸ Motors are used to rotate the washing in the drum.

▸▸ Motors can operate pumps in central heating systems to pump hot water around the radiators.

▸▸ Motors in computer-controlled greenhouses can turn watering systems on and off.

▸▸ Motors in computer-controlled greenhouses can open windows or close them to decrease or increase humidity.

▸▸ Motors in microwave ovens can rotate a turntable.

▸▸ Motors can control the movements of robots in factories.

▸▸ Motors can control the passage of items along a production line in a factory.

23

Buzzers

Buzzers often act as warning signs or signs to say a control process has been completed. For example, when a washing machine or drier has finished, the control system issues a buzz/beep to inform the householder. Buzzers can also be found in automatic cookers and microwaves.

Heaters

Heaters can be switched on/off according to control signals issued by a computer. The range of control devices with heaters is wide and here are some of them: automatic washing machines, automatic cookers, central heating controllers, and computer-controlled greenhouses.

Lights/lamps

Lights and lamps can be controlled using a control system that has a light sensor to sense the light level. In a computer-controlled greenhouse, optimum lighting conditions are maintained by switching on lights when the natural light levels fall below a certain value. Security lights and outside lighting can be controlled to come on when it goes dark and go off when it gets light again. Some cars also turn lights on and off automatically depending on light levels.

QUESTIONS B

1 Here are some input devices being used.

Give the name of the input device. *(5 marks)*

a

b

c

d

e

2 The manager of an office is buying a printer for the office. They are looking to buy one of the following types of printer:

Laser printer

Inkjet printer

Discuss the relative advantages and disadvantages of each of these types of printer. *(4 marks)*

REVISION QUESTIONS

1 Here is a list of devices that may be attached to a computer system:

LCD screen	keyboard
portable hard drive	mouse
touchpad	laser printer
flash/pen drive	microphone
digital camera	speakers
web camera	CD ROM drive

    a Write down the names of all the output devices in the above list. [3]

    b List **two** other output devices not in the above list. [2]

    c Give the name of **one** input device that is **not** included in the list above. [1]

    d Give the name of **one** device in the list above that could be used to back up data and programs. [1]

2 **a** Give **two** uses for each of the following input devices in a personal computer: [6]

       i mouse

      ii microphone

     iii digital camera

    b Give the names of **four** output devices and give **one** use for each of them. [8]

3 Copy and complete the table below. [10]

Application	Most suitable output device
Alerting the user that an error has occurred by making a beep	
Printing a poster in colour	
Listening to a radio station using the Internet	
Producing a large plan of a house	
Producing a hard copy of a spreadsheet	
Producing a colour picture on paper taken with a digital camera	
Producing a series of invoices with several copies that can be sent to different departments	
Producing a warning when a bar code is read incorrectly	
For listening to messages from a voicemail system	
Displaying the results of a quick search on the availability of a holiday	

4 Touch screens can often be seen at tourist information offices.

    a Describe what a touch screen is and how it works. [2]

    b Give **one** advantage of using a touch screen as an input device for use by the general public rather than using a monitor and a mouse. [1]

5 There are a number of different printers, each with their own advantages and disadvantages. The names of these printers are listed here:

    Laser printer

    Inkjet printer

Identify the name of the printer being described for each one of the following:

    a A printer which is used in offices for printing lots of documents in a short period of time. [1]

    b An inexpensive printer that is ideal for the home which can print in colour as well as black and white. [1]

    c A printer which sprays the ink onto the page. [1]

    d The type of printer where you have to be careful not to smudge the damp printouts as they come out the printer. [1]

    e The type of printer which uses a toner cartridge. [1]

    f The printer which is cheap to buy but which has high running costs owing to the high cost of the ink cartridges. [1]

Test yourself

Using the words in the list below, copy out and complete the sentences A to O underlining the words that you have inserted. Each word may be used more than once.

input   touchpad   keyboard   microphone

scanner   optical character recognition   laser

joysticks   output   sensors   mouse   stylus

inkjet   digital   touch

A   Devices used to get data from the outside world into the computer are called _____ devices.

B   The commonest input device, which comes with all computers, is the _____.

C   A _____ is used to move a pointer or cursor around the screen and to make selections when a desktop computer is used.

D   Where space is restricted, such as when a laptop is being used on your knee, a _____ is used instead of a mouse.

E   _____ are used primarily with games software.

F   In voice recognition systems a _____ is used as the input device.

G   The device used to input text and images is called a _____.

H   Special software can be used to recognize the individual letters in a scanned piece of text and this is called _____.

I   Cameras that do not use film and can transfer an image to the computer are called _____ cameras.

J   A pen-like device used to draw or write on a tablet is called a _____.

K   Quantities such as temperature and pressure can be detected and measured using _____.

L   Printers and plotters are examples of _____ devices.

M   The type of printer that is very fast and uses a toner cartridge is called a _____ printer.

N   A cheaper printer, which squirts a jet of ink at the paper, is called an _____ printer.

O   _____ screens are sometimes as an input device for multimedia systems and are popular for information points to be used by the general public.

EXAM AND EXAM-STYLE QUESTIONS

1 Ring **two** items which are output devices. *[2]*

| Blu-ray disc | Graph plotter | Graphics tablet |
| Optical mark reader | Projector | Web cam |

2 Draw **four** lines on the diagram to match the input device to its most appropriate use. *[5]*

Input device	Use
Scanner	For recording narration to be used with a presentation
Touch screen	For digitizing an old photograph so it can be put on a website
Chip reader	For inputting selections when buying a train ticket
Microphone	Reading information on a credit/debit card

3 Complete each sentence using **one** item from the list. *[4]*

| Laser | Dot matrix | Graph plotters | Projectors |
| Speakers | Inkjet | Actuators | Touch screens |

a .... printers are printers that use continuous stationery.

b .... are used when very large hard copy is required.

c .... are used with PDAs for both input and output.

d .... printers are used for the printing of high quality photographs.

4 A payroll office, which prints out large numbers of payslips every month, has decided to install a new printer. Discuss the advantages and disadvantages of using a laser printer, an inkjet printer or a dot matrix printer in this office. *[6]*

(Cambridge IGCSE Information and Communication Technology 0417/13 q20 Oct/Nov 2010)

5 Name the input devices **A**, **B**, **C** and **D** using the words from the list.

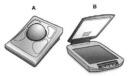

Chip reader	Digital camera	Joystick
Light pen	Microphone	Remote control
Scanner	Trackerball	

[4]

(Cambridge IGCSE Information and Communication Technology 0417/13 q1 Oct/Nov 2010)

6 Complete each sentence below using **one** item from the list.

A bar code reader	A dot matrix printer
A graph plotter	A graphics tablet	A microphone
A multimedia projector	A pressure sensor
A speaker	A touch pad	A web cam

a .... is used in CAD applications to produce very large printouts.

b .... is used to display data from a computer onto a very large screen.

c .... sends data to a microprocessor in a washing machine.

d .... is used to control the pointer on a laptop screen.

e .... is used to input sounds for use in presentation software. *[5]*

(Cambridge IGCSE Information and Communication Technology 0417/13 q4 Oct/Nov 2010)

3 Storage devices and media

Backing storage is any storage which is not classed as ROM or RAM. It is used to hold programs and data. Backing storage devices include magnetic hard drives, optical drives (CD or DVD), flash/pen drives, etc. Storage devices are the pieces of equipment which record data onto the storage media (i.e. optical, magnetic, etc.) or read it from the storage media.

In this chapter you will be covering both storage devices and media.

The key concepts covered in this chapter are:
▶▶ Description of the common backing storage media and their corresponding devices
▶▶ Identification of typical uses of the storage media, including types of access and access speeds
▶▶ Description of the comparative advantages and disadvantages of using different backing storage media
▶▶ Defining backups and explaining the reasons for taking them
▶▶ Description of the difference between main/internal memory and backing storage and the relative benefits of each

Storage devices and media

Data needs a material on which to be stored, which is called the storage media. For example, the storage media could be magnetic or optical or solid state in the case of a memory card. The storage device is the piece of equipment that is used to record the data onto the media or read it from the media.

Main internal memory and secondary memory

Main internal memory is of two types: ROM (read only memory) and RAM (random access memory).

ROM is:

▶▶ Permanent – so cannot be changed by the user.
▶▶ Non-volatile – meaning that the contents are not lost when the power is switched off.
▶▶ Used for boot routines – these are the sets of programs used to start the computer up when you turn the power on.

RAM is:

▶▶ Temporary – the contents are continually changing.
▶▶ Volatile – the contents are lost when the power is lost.
▶▶ Used to hold the operating system.
▶▶ Used to hold the software in use.
▶▶ Used to hold the files currently being worked on.

Secondary memory is storage other than ROM or RAM and is:

▶▶ Non-volatile, which means it holds its contents when the power is removed.
▶▶ Used to hold software/files not being used.

Ways of accessing data

There are two ways in which data is read from storage devices called:

▶▶ Serial /sequential access.
▶▶ Direct/random access.

○ **KEY WORD**

Storage media the collective name for the different types of storage materials such as DVD, magnetic hard disk, solid state memory card, etc.

Serial/sequential access

With serial/sequential access, data is accessed from the storage media by starting at the beginning of the media until the required data is found. It is the type of access used with magnetic tape and it is a very slow form of access when looking for particular data on a tape. It is, however, still used when every bit of data needs to be read off the tape and is used in batch processing systems where all the data on the tape is accessed and then processed in turn. For example, the preparation of utility bills (e.g. bills for water, gas, electricity) requires that all the data on the tape is read and processed in sequence to produce the bills for the customer.

In order to update a magnetic tape with the changes made it is necessary to merge some of the information from the old tape with the new information which is then written to a completely new tape.

Direct/random access

With direct/random access, data is accessed immediately from the storage media. This is the method used with storage media such as magnetic hard disks and optical media such as CD and DVD. It is used when fast access to the data is important such as online booking systems, systems for point-of-sale (POS) terminals in shops, etc.

With direct/random access the computer knows where all the data is stored on the disk so it can find any data extremely quickly. It also means that the new data produced during updating can be stored anywhere on the disk.

Fixed hard disks

Fixed hard disks consist of a series of disks coated with a magnetic material and a series of read/write heads which record data onto the surface or read it off the surface. Fixed hard disks are used to store operating systems, applications software (i.e., the software you use to complete tasks such as word-processing software), and any files of data needed by the user. All of these require that the data can be both accessed and stored onto the media quickly.

Fixed hard disks can consist of one or more magnetically coated disks.

Fixed hard disks are, as the name suggests, fixed in the computer and are not designed to be portable. Portable hard disks are available and these will be looked at later.

Fixed hard disks in PCs and laptops are used for storage of files that have been created using software and then stored, e.g. c:/My Documents/letter.doc.

Fixed hard disks are used for online processes. For example, when you create a personal web site, it is stored on the fixed hard disk of the organization you use to connect to the Internet. When other people access your web site, the data is obtained off this fixed hard disk. Computers connected to the Internet which store web pages are called web servers.

Many fixed hard disks are used in file servers for networks. In most networks all the data and program files are stored centrally rather than on each individual computer. This central store of data uses fixed hard disks as the storage media. Because many users will want access to this store, the speed at which data is accessed and transferred needs to be very high.

Advantages of fixed hard disks:

▸ A very high access speed (i.e., speed of finding and reading data off the disk).
▸ A very high transfer rate (i.e., speed of storing data onto the disk).
▸ A very high storage capacity.

Disadvantages of fixed hard disks:

▸ Hard disks cannot be transferred between computers unlike a portable hard disk.
▸ The hard disk cannot be taken out of the computer and locked away for security purposes.

Portable hard disks

It is possible to buy additional hard disks. These hard drives are called portable hard disks and may be removed each night and stored safely.

Portable hard disks are also used to store very large files which need to be transported from one computer to another. Generally portable hard disks are more expensive than other forms of removable media but their very large storage capacity, high access speed, and transfer rate are the reasons why they are chosen. Portable hard disks are used for reasons other than the back-up of data and programs, e.g. a writer who works in two or three locations during the week.

Most portable hard disks connect to a USB port and need no installation. You simply connect them and the operating system knows how to work them.

Advantages of portable hard drives include:

▸ Their very large storage capacity means large files can be transferred between computers.
▸ They are very fast at accessing and storing files so transferring large files such as multimedia files takes little time.
▸ They can be attached to and used by any computer that has a USB port.

Disadvantages of portable hard drives include:

▸ Their size and portability means they are easily stolen.
▸ It makes it easy for employees to copy confidential data such as a medical database which is a security risk to companies/organizations.
▸ Their use with lots of different computers can mean there is a danger of viruses being transferred.

Magnetic tapes

Magnetic tape stores data on a plastic tape coated with a magnetic layer. Here are the main features of magnetic tape:

▸ Magnetic tape has a huge storage capacity.
▸ Magnetic tape is used to back up the data stored on hard disks.
▸ Because it takes time to move the tape to the position where the data is stored, tape storage is much less common than disk storage.

Magnetic tape is used in any application where extremely large storage capacity is needed and the speed of access is not important.

Magnetic tape has a huge storage capacity but is only used for a few limited applications.

Magnetic tape provides serial access. What this means is that it is necessary to access each record in turn on the tape until the correct file is found. This takes a long time and is the reason why magnetic tape is being taken over by fixed hard disks. Magnetic tape is useful for when every record on the tape needs to be accessed or stored in turn on the tape. This is the reason why magnetic tape can be used for backups of file servers for computer networks. These servers can have huge storage capacities and can be backed up on a single tape rather than a series of hard disks.

Magnetic tape is also used in a variety of batch processing applications where the computer can just be left to process the data without any human intervention during the processing. Examples of this include reading bank cheques, payroll processing and general stock control.

Advantages of magnetic tape include:

▸▸ They have extremely large storage capacities and this makes them ideal for backup purposes where all data and programs need to be stored.
▸▸ They are less expensive compared to similar capacity magnetic hard disk drives.
▸▸ The data transfer rate is high (writing to tape).

These magnetic tape cartridges are used for daily backups of data.

Disadvantages of magnetic tape include:

▸▸ They are not suitable for an application that requires fast access to data because the speed of access is low.
▸▸ In order to update details on a tape it is necessary to create a new tape containing some of the previous details along with the updated details.

Optical drives

Optical disks are flat circular disks on which data is stored as a series of bumps. The way the bumps reflect laser beam light is used to read the data off the disk.

CDs are used to hold large files (<1 gigabyte) and are ideal for holding music and animation files. DVDs have a much larger capacity 4.7 to 8.5 gigabytes and are used mainly for storing films/video. Both CD and DVD can be used to store computer data and can be used for backup purposes.

Optical disks include CDs and DVDs and are used to store digital data as a binary pattern on the disks.

CD ROM (compact disk read only memory)

CD ROMs are used mainly for the distribution of software and the distribution of music. Although most home computers are equipped with DVD drives, a lot more computers, especially those used in businesses, still only have CD drives. You can read a CD using a DVD drive but you cannot read a DVD with a CD drive. This is why software is still being sold on CD rather than DVD.

With CD ROM:

▸▸ data is read only
▸▸ data is stored as an optical pattern
▸▸ there is a large storage capacity (700 Mb)
▸▸ they can be used for the distribution of software.

Advantage:

▸▸ Once written, the data cannot be erased and this makes it useful for the distribution of software/music or backing up.

Disadvantage:

▸▸ Data transfer rate and access rate are lower than for a hard disk.

DVD ROM (digital versatile disk read only memory)

DVD ROM is used for the distribution of movies where you can only read the data off the disk. A DVD ROM drive can also be used for reading data off a CD. DVD is mainly used for the distribution of films and multimedia encyclopaedias.

Advantage:

▶▶ High storage capacity means full-length movies can be stored.

Disadvantages:

▶▶ The user cannot store their files on the disk.
▶▶ Older computers may not have a drive capable of reading DVDs.

CD R (CD recordable)/DVD R (DVD recordable)

CD R allows data to be stored on a CD, but only once. DVD R allows data to be stored on a DVD once. Both these disks are ideal where there is a single "burning" of data onto the disk. For example, music downloaded off the Internet could be recorded onto a CD in case the original files were damaged or lost.

They can be used for archive versions of data. Archive versions are where old data is stored in case it is needed in the future. Storing archive data on the fixed hard disk would clutter up the disk so it is better to store it on removable media and store it in a safe place. DVD R is ideal for storing TV programmes where you do not want to record over them.

CD RW (CD rewriteable)

A CD RW disk allows data to be stored on the disk over and over again – just like a hard disk. This is needed if the data stored on the disk needs to be updated. You can treat a CD RW like a hard drive but the transfer rate is less and the time taken to locate a file is greater. The media is not as robust as a hard drive.

Advantages:

▶▶ Re-writable so can be re-used.
▶▶ The data stored can be altered.

Disadvantages:

▶▶ The data transfer rate is lower than for a magnetic hard disk.
▶▶ Optical drives such as CD RW are more easily damaged than hard drives.

DVD RW (digital versatile disk read/write)

A DVD RW drive can be used to write to as well as read data from a DVD. DVD RW are sometimes called DVD burners because they are able to be written to and not just read from. Like CD RW, they are ideal for storing data that needs regularly updating.

Typical storage capacities are:

▶▶ 4.7 Gb for the older DVD drives.
▶▶ 8.5 Gb for the latest DVD drives.

DVD RAM (digital versatile disk random access memory)

DVD RAM has the same properties as DVD RW in that you can record data onto it many times but it is faster and it is easier to overwrite the data. The repeated storage and erasure of data acts in a similar way to RAM – hence the name. A typical storage capacity for DVD RAM is 10 Gb. DVD RAM is used for the storage of TV/film at the same time as watching another.

Advantages:

▶▶ Fast transfer rate
▶▶ Fast access rate.

Disadvantage:

▶▶ The devices are expensive compared to other devices.

Blu-ray

The Blu-ray disk is a new optical disk that has a much higher storage capacity than a DVD. Blu-ray disks have capacities of 25 Gb, 50 Gb, and 100 Gb. These high capacity Blu-ray disks are used to store high definition video. They are used for storing films/movies with a 25 Gb Blu-ray disk being able to store 2 hours of HDTV or 13 hours of standard definition TV. It is possible to play back video on a Blu-ray disk whilst simultaneously recording HD video. Newer computers now come with Blu-ray drives and eventually Blu-ray disks will become the norm for the storage of data on computers.

The high capacity of Blu-ray makes them suitable for storing HDTV.

Solid state backing storage

Solid state backing storage is the smallest form of memory and is used as removable storage. Because there are no moving parts and no removable media to damage, this type of storage is very robust. The data stored on solid state backing storage is rewritable and does not need electricity to keep the data. Solid state backing storage includes the following:

▶▶ memory sticks/pen drives
▶▶ flash memory cards.

> **! Revision Tip**
>
> When selecting backing storage media, ensure that you take into account the relative access speeds, whether the media needs to be re-used and the size of the files to be stored.

Memory sticks/pen drives

Memory sticks/pen drives are very popular storage media which offer large storage capacities and are ideal media for photographs, music, and other data files. Memory sticks are more expensive per Gb than CD/DVD/hard disk. They consist of printed circuit boards enclosed in a plastic case.

The main advantages are:

▸▸ Small and lightweight – easy to put on your key ring or in your pocket.
▸▸ Can be used in almost any computer with a USB drive.
▸▸ Large storage capacity (up to 256 Gb).
▸▸ No moving parts so they are very reliable.
▸▸ Not subject to scratches like optical media.

The main disadvantages are:

▸▸ Their small size means they are easily stolen.
▸▸ They are often left in the computer by mistake and lost.
▸▸ They have lower access speeds than a hard disk.

Memory sticks/pen drives are the most popular portable storage media. Their portability is their main advantage and you simply plug them into the USB port where they are recognized by the operating system automatically.

Memory sticks/pen drives are ideal for the transfer of relatively small amounts of data between computers.

Flash memory cards

Flash memory cards are the small thin rectangular or square removable cards that are used for storage of digital images by digital cameras. They can also be used in any situation where data needs to be stored and so are used with desktop computers, laptops, palmtops, mobile phones, and MP3 players. You can see the card readers in supermarkets and other stores where you can take your cards containing photographs and get them printed out.

This flash memory card is being used with a mobile phone.

Backups and the reasons for taking them

A backup is a copy of data and program files kept for security reasons. Should the originals be destroyed or corrupted then the backups can be used. Using a file server and storing both programs and data on it, means that backups can be taken in one place. Backups should be held on removable devices or media that are taken off site each day. The individual users do not need to take their own backups. The person in charge of the network (i.e., usually the network manager) will take the backups needed. Many systems now take backups automatically at a certain time of the day and send the data using the Internet to a company that specializes in storing backups. Backups should always be removed off-site. This is in case of fire or if the building is destroyed.

QUESTIONS A

1 Which **three** of the following are backing storage devices? *(3 marks)*
> RAM
> Hard drive
> CD RW drive
> ROM
> Plotter
> Speaker
> Keyboard
> Pen drive

2 Give the meaning of the following abbreviations.
   a DVD R *(1 mark)*
   b DVD RW *(1 mark)*

3 Backups of programs and data should be taken on a regular basis.
   a Explain what is meant by a backup. *(1 mark)*
   b Give **one** reason why backups should be taken on a regular basis. *(1 mark)*
   c Give **one** example of backing storage suitable for the taking of backup copies and explain why it is suitable. *(3 marks)*

Activity 3.1

Everyone who uses a computer needs to store their data and programs somewhere.

For this activity you have to find out about the backing storage devices that are available for computer users.

Use the Internet to find out:

▸▸ Types of storage devices and their storage capacity.
▸▸ The main advantages and disadvantages of each device.
▸▸ Type of access.
▸▸ Speed of access.
▸▸ The cost of the storage device.
▸▸ The media they need if applicable.

A good place to look for information is the web site of the large online stores that sell computer equipment.

Test yourself

The following notes summarize this topic. The notes are incomplete because they have words missing. Using the words in the list below, copy out and complete the sentences A to G, underlining the words that you have inserted. Each word may be used more than once.

cartridges   immediately   DVD
backing   hard   programs   backups

A _____ are copies of data and program files kept for security reasons.

B Backing storage is used for the storage of programs and data that are not needed _____ by the computer.

C Flash/pen drives are the most popular _____ storage media because they are very small and cheap compared to the alternatives.

D Currently they are also used for storing backups of _____ and data in case the originals are damaged or destroyed.

E _____ drives consist of a series of disks with a magnetic coating and a series of read/write heads which put the data onto or record it off each surface.

F _____ are usually use reels of magnetic tape in a hard plastic case.

G Optical media include CD and _____ for storing data.

REVISION QUESTIONS

1 a Give **two** uses for each of the following devices in a personal computer:
  i Hard disk drive   [2]
  ii CD ROM drive   [2]
  b Give **two** ways small high capacity storage devices have influenced the development of portable equipment that can be used for work and play.   [2]

2 Computers need memory and backing storage.
  a Give **one** difference between main internal memory and backing storage.   [1]
  b Give **one** example of what would be stored in main internal memory.   [1]
  c Give **one** example of what would be stored in backing storage.   [1]

3 Data needs to be stored for future use. Here are a number of storage devices/media. For each of these, explain a suitable use and explain clearly why the storage device/media is suited to the application.
  a Memory card   [3]
  b CD ROM   [3]
  c Magnetic hard drive   [3]
  d Flash/pen drive   [3]

EXAM AND EXAM-STYLE QUESTIONS

1 Ring **two** items which are storage media.

Flash memory card   Graph plotter   Magnetic disc

OCR   OMR   Touch pad

       [2]

(Cambridge IGCSE Information and Communication Technology 0417/13 q2 Oct/Nov 2010)

2 A student wants to transfer work from a computer in school to their home computer.
They have the choice of using a CD or a pen drive.
Give **three** reasons why a student would choose a pen drive rather than a CD.   [3]

3 Name the methods of storage **A**, **B**, **C** and **D** using the words from the list.

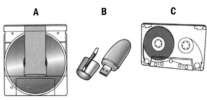

    A   **B**   **C**   **D**

Bar code   Chip   DVD RAM

Graphics tablet   Light pen   Magnetic disc

Magnetic tape   Pen drive   [4]

(Cambridge IGCSE Information and Communication Technology 0417/11 q1 May/June 2010)

4 A website designer has the choice of backing their work up using either a pen drive or a portable hard disk.
Give **two** advantages of using the portable hard disk for backup.   [2]

4 Computer networks

Most computers are now connected to networks. For example, in the home your computer may be able to access the Internet, in which case it becomes part of a network. You may have a small network at home, which allows you to access the Internet on a desktop computer and a laptop at the same time. Networks provide so many more benefits compared to stand-alone computers (i.e., computers not connected to a network).

The key concepts covered in this chapter are:
▸▸ The key components of networks and their purposes
▸▸ The advantages and disadvantages of networking
▸▸ The use of Wi-Fi and Bluetooth
▸▸ Communication methods
▸▸ The differences between local area networks (LANs) and wide area networks (WANs)
▸▸ LANs, WANs and WLANs
▸▸ Common network environments such as intranets and the Internet
▸▸ Problems of confidentiality and security of data in network environments
▸▸ Setting up small networks

What is a network?

A network is two or more computers that are linked together so that they are able to share resources. These resources could be a printer, scanner, software or even a connection to the Internet. You can also share data using a network. For example, a pupil database in a school could be accessed from any of the computers connected to the network.

The advantages and disadvantages of using computer networks

Advantages:

▸▸ You can share hardware – you share printers, scanners and the equipment such as modems and routers used to provide an Internet connection.
▸▸ Software is allocated out to each computer from a central position or hosted. This makes the software faster to install as the software only needs installing once. The software is easier to maintain as only one copy of the software is updated. A network copy of the software is cheaper to purchase than individual copies for stand-alone computers.
▸▸ Work can be backed up centrally by the network manager, which means users do not have to back up their own work. The network manager will make sure that the work is backed up.
▸▸ Passwords make sure that other people cannot access your work unless you want them to.
▸▸ Speed – it is very quick to copy and transfer files.
▸▸ Email facilities – any user of the network will be able to communicate using electronic mail. This will be much more efficient compared to paper-based documents such as memos, etc.
▸▸ Access to a central store of data – users will have access to centrally stored data.

Disadvantages:

▸▸ A network manager may need to be employed – this can be quite expensive.

▸▸ Security problems – viruses are spread more easily. Hackers may also gain access to the network.
▸▸ Breakdown problems – if the network breaks down, users will not have access to the important information held.
▸▸ Initial costs of equipment and cabling – a server and cables and/or other communication devices will be needed. The installation costs of a network are also high, particularly sinking cables into walls, ceilings, etc.

The two types of network: LAN and WAN

There are two types of network: a local area network (LAN) and a wide area network (WAN).

Basically a WAN is much bigger than a LAN and is spread over different sites. A LAN, however, is within one site or building. This table gives you the main features of each type of network.

LAN	WAN
Confined to a small area	Covers a wide geographical area (e.g. between cities, countries and even continents)
Usually located in a single building	In lots of different buildings and cities, countries, etc.
Uses cable, wireless, infra-red and microwave links which are usually owned by the organization	Uses more expensive telecommunication links that are supplied by telecommunication companies (e.g. satellite links)
Less expensive to build as equipment is owned by the organization which is cheaper than renting lines and equipment	More expensive to build as sophisticated communication systems are used involving rental of communication lines

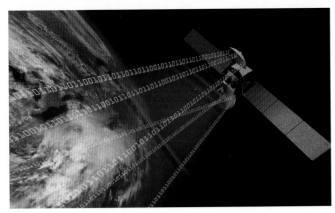

Wide area networks often make use of satellite links.

KEY WORDS

Network a group of computers that are able to communicate with each other.
LAN (local area network) a network of computers on one site.
WAN (wide area network) a network where the terminals/computers are remote from each other and telecommunications are used to communicate between them.

Network topologies

The devices in a network may be arranged in different ways. Each way is called a topology. It is important to note that in a wired network the topology would show how the wires are connected. However, many networks are now set up without wires, making use of radio, infra-red or satellite links. The topologies in this case will show the communication links between the devices. The main topologies are:

- bus
- star
- tree.

The bus topology

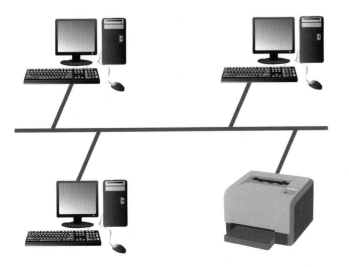

With a bus topology:

- All the devices connected to the network are connected to a common shared cable called the backbone/central line.
- Signals are passed in either direction along the backbone.

Advantages of bus topology networks:

- Lower cost because less cabling is needed.
- Simple cable runs makes them easy to install.
- It is easy to add extra devices to the network.
- If one cable (other than the backbone) breaks, then only that PC is affected.

Disadvantages of bus topology networks:

- If more than about 12 devices are connected to the network, then the performance of the network is poor.
- If there is a break in the backbone cable, then the whole network fails.

Star topology

The star topology uses a central connection point for all the devices on the network. The central connection point can be a hub or switch.

Advantages of star topology networks:

- Faults in network cables will not affect the whole network – if one of the cables fails, then the other computers on the network can still be used.
- Easy to add more computers – extra computers can be added without much loss in performance because all computers have their own path to the hub or switch.

Disadvantages of star topology networks:

- Higher cost – the large amount of cabling needed makes it a more expensive topology.
- Dependence on the central hub/switch – if the device at the centre of the network fails, then the whole network will fail.

Tree topology

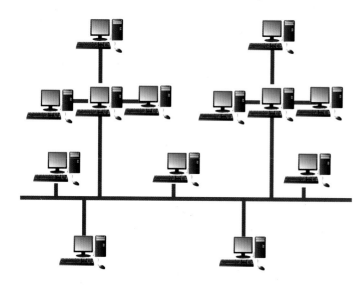

A tree topology is a combination of a bus topology and a star topology. Look at the diagram and notice the way the two star networks are connected to the bus (shown here as a horizontal cable). If the bus cable failed, the computers in a star topology network would communicate with each other but there could be no communication from one star to other stars.

Advantages of tree topology networks:

▶▶ All the computers have access to their immediate network as well as the larger network.
▶▶ Good for networks that are widely spread and have lots of branches.

Disadvantage of tree topology networks:

▶▶ Not suitable for small networks as it wastes cable.

WLAN

WLAN stands for wireless local area network and means a LAN where the computers are able to communicate with each other wirelessly. WLAN is the type of network you have in your home where several computers are all able to access the Internet at the same time.

A WLAN allows users to move around an area and yet still stay connected to the Internet.

The use of Wi-Fi and Bluetooth

There are two main technologies that allow communication between computers and other devices wirelessly.

Wi-Fi

Wi-Fi enables computers and other devices to communicate wirelessly with each other. Areas where the Internet can be accessed wirelessly using Wi-Fi are called access points or hotspots and they can be found in many public places and coffee bars, hotels, etc.

The range of a WLAN depends on the wireless access point (WAP) or the type of wireless router being used. It also depends on if there are obstacles such as walls in the way of the signal. For a home network, the distances are typically 50m indoors and 100m outdoors.

Advantages of wireless communication:

▶▶ Provided you have a wireless signal, you can work in hotels, outside, in coffee shops, etc.
▶▶ You are not confined to working in the same place. For example, you can work on trains, buses and even some aircraft provided there is a signal.
▶▶ Fewer/no trailing wires to trip over.
▶▶ It is easier to keep a working area clean if there are not as many wires in the way.
▶▶ There are no network wires so there are no costs for their installation.

Disadvantages of wireless communication:

▶▶ The danger of hackers reading messages.
▶▶ There are areas where you cannot get a wireless network.
▶▶ There is some evidence that there may be a danger to health.
▶▶ Limited signal range.

Bluetooth

Bluetooth is a wireless technology used to exchange data over short distances and it makes use of radio waves. The range of Bluetooth depends on the power of the signal and can be typically from 5m to 100m.

Here are some uses for Bluetooth:

▶▶ Wireless communication between input and output devices. If you have a wireless keyboard and mouse then they could use Bluetooth. Printers can be controlled wirelessly from a computer using Bluetooth.
▶▶ Communication between a wireless hands-free headset and a mobile phone. These are the sort people can use whilst they are driving.
▶▶ Creating a small wireless network where the computers/devices are near to each other and where the small bandwidth is not a problem. For example, creating a small home wireless network.
▶▶ Transferring appointments, contacts, etc., between a computer and a mobile device such as mobile phone or PDA.
▶▶ Communication using a games controller. Many games consoles use Bluetooth.

Bluetooth is used with hands-free headsets.

The Internet

The Internet is a huge group of networks joined together. Each of these networks consists of lots of smaller networks. When you connect to the Internet your computer becomes part of the largest network in the world.

The advantages of the Internet:

- Huge amounts of information can be accessed almost anywhere.
- Improved communication systems – this includes the use of text messages, emails, instant messaging, etc.
- Changes in the way we shop – many people prefer to shop online.
- VoIP (Voice over Internet Protocol) – enables cheap international phone calls to be made using the Internet rather than having to pay for calls using a mobile or landline telephone. VoIP allows voice data to be transferred over the Internet and allows phone calls to be made without the need to subscribe and pay for a phone service.
- Can help people with disabilities to be more independent – because they can order goods and services online.

The disadvantages of the Internet:

- Misinformation – there are many bogus or fake sites so you need to check the information you obtain carefully or use only reliable sites.
- Cyber crime – we have to be much more careful about revealing personal information such as bank and credit card details.
- Addiction to gambling, as there are many casino, bingo, horse racing, etc., betting sites.
- Increased problems due to hacking and viruses.
- Deserted city centres as shops close down since they cannot compete with Internet shopping.
- Paedophiles look for children using the Internet.

There are a number of devices and systems that you need to access the Internet and these are outlined here.

E-commerce

Most organizations use websites sometimes for promotional purposes and sometimes to allow people to order goods or services using the site. This is called e-commerce.

QUESTIONS A

1 Give **two** benefits of using a local area network. *(2 marks)*

2 Give **one** difference between a local area network (LAN) and a wide area network (WAN). *(1 mark)*

3 Describe **two** advantages and **two** disadvantages in a company using a computer network. *(4 marks)*

4 The computers in a network are connected in a topology.
  a Explain what is meant by a network topology. *(2 marks)*
  b Give the names of **two** different network topologies. *(2 marks)*

5 Bluetooth is a method that allows devices to communicate with each other and pass data.
  a Give the names of **two** pairs of devices that can communicate using Bluetooth. *(2 marks)*
  b Explain **one** advantage in devices communicating using Bluetooth. *(1 mark)*
  c Explain **one** disadvantage in devices communicating using Bluetooth. *(1 mark))*

6 An office is thinking of introducing a wireless network with a wireless connection to the Internet.
  Give **two** advantages in using a wireless network rather than a wired one. *(2 marks)*

7 A small office uses a local area network. The telecommunications company supplies them with a cable into the building. They would like to set up a wireless network.
  a Give **one** reason they would like their network to be wireless. *(1 mark)*
  b A wireless router is bought. Give the purpose of a wireless router. *(1 mark)*

8 A company has a wireless network installed. Give **one** reason why they might be concerned about the security of their data. *(1 mark)*

⊙ KEY WORDS

ISP (Internet service provider) a company that provides a connection to the Internet.

Web/Internet portal a website that acts as a point of access to the World Wide Web. Includes a search engine and access to other services such as email, news, stock prices, weather, entertainment, etc.

Many of these websites allow customers to browse online catalogues and add goods to their virtual shopping basket/trolley just like in a real store. When they have selected the goods, they go to the checkout where they have to decide on the payment method. They also have to enter some details such as their name and address and other contact details. The payment is authorized and the ordering process is completed. All that is left is for the customer to wait for delivery of their goods.

The difference between Internet and WWW

There is a difference between the Internet and the World Wide Web (WWW). The Internet is the huge networks of networks. It connects millions of computers globally and allows them to communicate with each other. The World Wide Web (WWW) is the way of accessing the information on all these networked computers and makes use of web pages and web browsers to store and access the information.

Intranets

An intranet is a network that is used inside an organization and makes use of web pages, browsers and other technology just like the Internet. Schools and colleges use intranets and they can hold all sorts of information from teaching resources, information about courses, and adverts to student personal records and attendance details. Parts of the intranet can be made available to anyone in the organization, while parts that contain personal details can be made available only to certain people. Restriction to certain parts of the intranet is achieved by using user-IDs and passwords.

The differences between an intranet and the Internet

» Internet stands for INTERnational NETwork, whereas intranet stands for INTernal Restricted Access NETwork.
» An intranet contains only information concerning a particular organization that has set it up, whereas the Internet contains information about everything.
» Intranets are usually only used by the employees of a particular organization, whereas the Internet can be used by anyone.
» Intranets are based on an internal network, whereas the Internet spans countries around the world.
» With an intranet, you can block sites which are outside the internal network using a proxy server.
» The information for an intranet is usually stored on servers.
» Intranets are usually behind a firewall, which prevents them from being accessed by hackers.
» An intranet can be accessed from anywhere with correct authentication.

The advantages in using an intranet:

» Intranets are ideal in schools because they can be used to prevent students from accessing unwanted information.
» The internal email system is more secure compared to sending emails using the Internet.
» Only information that is relevant to the organization can be accessed and this saves employees accessing sites that are inappropriate or which will cause them to waste time.

ISPs and the services they provide

Connecting directly to the Internet is very expensive and only suitable for large companies and organizations. Computers used to supply an Internet connection for other computers are called web servers. Most people connect to the Internet via an organization called an Internet service provider, or ISP for short. This is an organization that supplies the connection to the Internet as well as providing services including:

» storage on their server, where you can store your website
» email facilities
» instant messages where you can send short messages to your friends when they are online
» access to online shopping
» access to news, sport, weather, financial pages, etc.

Web browser software

Web browser software is a program that allows web pages stored on the Internet to be viewed. Web browsers read the instructions on how to display the items on a web page which are written in a form called HTML (Hypertext Markup Language). You will be learning about HTML in Chapter 15. A web browser allows the user to find information on web sites and web pages quickly and it does this through:

» entering a web address (i.e. URL)
» a web/internet portal
» key word searches
» links
» menus.

Web browser software includes email software that allows you to send and receive email.

Bridges

Bridges are used to connect LANs together. When one of the LANs sends a message, all the devices on the LAN receive the message. This increases the amount of data flowing on the LAN. Often a large LAN is divided into a series of smaller LANs. If a message is sent from one computer in a LAN to another computer in a different LAN then the message needs to pass between the LANs using a bridge. The advantage in subdividing a larger network is that it reduces the total network traffic as only traffic with a different LAN as its destination will cross over the bridge. A bridge therefore usually has only two ports in order to connect one LAN to another LAN.

Routers

Each computer linked to the Internet is given a number which is called its IP (Internet Protocol) address. This address is like this 123.456.1.98 and is unique for each device whilst linked to the Internet.

When data is transferred from one network to another the data is put into packets. The packets contain details of the destination address of the network it is intended for. Computers on the same

QUESTIONS B

1 Explain the difference between the terms Internet and intranet. *(2 marks)*

2 Home users usually connect to the Internet using an ISP.
   a State the meaning of ISP. *(1 mark)*
   b As well as providing a connection to the Internet, an ISP provides other services. Describe **two** other services provided by an ISP. *(2 marks)*

3 Describe what is meant by web browser software. *(2 marks)*

network all have the same first part of the Internet Protocol address and this is used to locate a particular network.

Routers are hardware devices that read the address information to determine the final destination of the packet. From details stored in a table in the router, the router can direct the packet onto the next network on its journey. The data packet is then received by routers on other networks and sent on its way until finally ending up at the final destination network.

Routers can be used to join several wired or wireless networks together. Routers are usually a combination of hardware which acts as gateways so that computer networks can be connected to the Internet using a single connection.

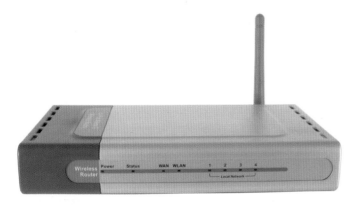

This wireless router allows all the computers in the home to share an Internet connection.

Common network devices

There are a number of other devices that you might find in some networks and these are outlined here.

Switches

Like hubs, switches are used to join computers and other devices together in a network but they work in a more intelligent way compared to hubs. Switches are able to inspect packets of data so that they are forwarded appropriately to the correct computer. Because a switch only sends a packet of data to the computer

it is intended for, it reduces the amount of data on the network, thus speeding up the network.

Hubs

A hub is a simple device that does not manage any of the data traffic through it. It is simply used in a network to enable computers on the network to share files and hardware such as scanners and printers.

Data is transferred through networks in packets. A hub contains multiple **ports** (i.e. connection points). When a **packet** arrives at one port, it is **transferred** to the other ports so that all network devices of the LAN can see all packets. Every device on the network will receive the packet of data, which it will inspect to see if it is relevant or not.

Proxy server

A proxy server is a server that takes client computer requests and forwards them to other servers. The proxy server sits between the server of the network you are using and the server of the system outside of your network that you need to access. The proxy server is able to modify requests and responses. A proxy server can be a major component of a firewall.

To understand proxy servers properly you need to understand how they are used. Here are some uses:

▸▸ Organizations such as schools, libraries and businesses use proxy servers to block offensive web content such as pornography.

▸▸ They can be used to protect anonymity on the Internet. Every server connected to the Internet will have an IP address. Once this is known a number of problems can occur such as hacking. Using the proxy server means that the main server for the network being protected is anonymous to all the other servers on the Internet.

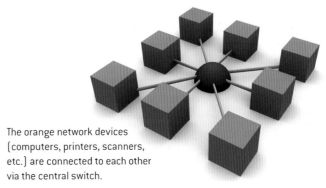

The orange network devices (computers, printers, scanners, etc.) are connected to each other via the central switch.

Problems of confidentiality of data

Using computer networks allows data to be transferred easily from one place to another. Some of this data is personal data and contains details such as credit card numbers, bank account details, medical information, etc.

The Internet is not normally secure, so special measures need to be taken to ensure that there is no unauthorized access to these details. The methods used include:

- usernames/user-IDs and passwords
- biometrics
- digital signatures
- encryption
- access levels – this means that only certain staff can access certain data
- training of staff – to ensure they understand the risks to personal data
- physical security of room.

Usernames and passwords are an essential part of network security.

Authentication techniques

Authentication techniques are those techniques that are used to check that a person accessing a network or communications system is the genuine person. They can also be used to ensure that an email sent by a person is genuinely from them and not somebody pretending to be them.

The main authentication techniques are:

- usernames and passwords
- digital signatures
- biometrics.

Identifying the user to the system: usernames

A username is series of characters that is used to identify a certain user to the network. A username must be unique, meaning that no two users will have the same username. The person who looks after the network will use this to allocate space on the network for the user. It is also used by the network to give the user access to certain files. The network manager can also keep track of what files the user is using for security reasons.

Preventing unauthorized access to the system: the use of passwords

A password is a string of characters (letters, numbers, and punctuation marks) that the user selects. Only the user will know what the password is. When the user enters the password, it will not be shown on the screen. Only on entry of the correct password will the user be allowed access to the network.

Characteristics of good passwords

A good password that will keep out hackers should have the following characteristics:

- Do not use a word that can be found in a dictionary.
- Do not use your own or any other name or surname even if you put numbers after it.
- Always use the maximum number of characters the system allows. So, if the password can be up to 10 characters, use all 10 and not just say 4 of them.
- Include numerals as well as letters.
- Include a mixture of upper and lower case letters but try not to make the first letter a capital letter.
- If the password system allows, put other characters in your password like £, &, %, $, @, etc.

When using passwords:

- Do not write your password down.
- Change your password regularly (usually the system will prompt you to do this automatically).
- Do not tell anyone else your password.
- Do not use your user-ID as your password.
- Never respond to an email that asks you for your password.

Biometrics

Biometrics uses features of the human body that are unique to a particular person. Usually this is a person's fingerprint or the pattern on the back of the eye (called the retina). Instead of logging in using passwords and username the user simply puts their finger into a scanner. The system automatically identifies their unique fingerprint and allows them access to the network and the resources they are allowed to use.

With retinal scanning, the user looks into the scanner eyepiece. A beam of infra-red light scans over the retina and converts the pattern of the reflected light into computer code. This code is used to find a match to details stored in a database. This authenticates the user.

Advantages of biometrics:

- You do not have to remember passwords and usernames.
- The system is hard to abuse unlike swipe card systems where the card could be used by someone else.

Disadvantages of biometrics:

- Some people worry about the privacy of data stored – especially fingerprint details.
- The readers are quite expensive.

Digital signatures

Ordinary signatures can be used to check the authenticity of a document. By comparing a signature you can determine whether a document is authentic. When conducting business over the Internet you also need to be sure that emails and electronic documents are authentic. Digital signatures are a method used with emails to ensure that they are actually from the person they say they are from and not from someone pretending to be them.

A fingerprint scanner being used.

"**Encryption software is expensive...so we just rearranged all the letters on your keyboard.**"

⊙ KEY WORD

Digital signature a way of ensuring that an email or document sent electronically is authentic. It can be used to detect a forged document.

Encryption

Because so many people use a network it is important to ensure that the system is secure. If there is access to the Internet via the network, then there needs to be protection against hackers. These hackers could simply intercept and read email or they could alter data or collect personal details and credit card numbers to commit fraud.

Encryption is used to protect data from prying eyes by scrambling data as it travels over the Internet. Encryption is also used when saving personal data onto laptops or removable storage devices. If any of these gets lost or stolen then the data cannot be read. Only the authorized person being sent the information will have the decryption key that allows them to unscramble the information.

Encryption should be used for the following:

▶▶ Sending credit card details such as card numbers, expiry dates, etc., over the Internet.
▶▶ Online banking.
▶▶ Sending payment details (bank details such as sort code numbers, account numbers, etc.).
▶▶ Confidential emails (e.g. with personal or medical details).
▶▶ Sending data between computers on a network where confidentiality is essential (e.g. legal or medical).
▶▶ Storing sensitive personal information on laptops and portable devices and media.

⊙ KEY WORDS

Encryption the process of scrambling files before they are sent over a network to protect them from hackers. Also the process of scrambling files stored on a computer so that if the computer is stolen, they cannot be read.
Password a series of characters chosen by the user that are used to check the identity of the user when they require access to an ICT system.
Username or user-ID a name or number that is used to identify a certain user of the network or system.

Methods of communication

Networking computers and other devices allow improved communication between people using services such as:

▶▶ fax
▶▶ email
▶▶ teleconferencing/videoconferencing.

Fax

Fax is like a long distance photocopier. You put the document in the fax machine at one end and then enter the telephone number of the fax machine to which it is to be sent. The page is scanned and converted to a bit map (binary pattern of the page), passed along the telephone line and it is printed out on the recipient's fax machine. Documents as well as diagrams/drawings can be sent by fax. Fax is rapidly being replaced by email as much of the information such as letters, contracts, plans, etc., is already in digital form so it can be attached to an email.

It is possible to use a computer with a scanner and a printer as a fax machine and instead of printing the fax out it can be saved and viewed on the screen.

Software can also be used to convert files and send them directly to a fax machine.

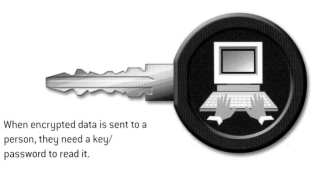

When encrypted data is sent to a person, they need a key/ password to read it.

A fax machine

Advantages in using fax:

- Emails can be used to distribute viruses. Fax does not spread viruses.
- Fax cannot be hacked into.
- Faxed signed documents are legally acceptable.

Disadvantages in using fax:

- Faxes are usually printed out in an area shared by other employees so they can be seen by others.
- Paper jams cause wasted time.
- The receiver's fax machine may not be turned on.
- The receiver's fax machine may be out of paper.
- Information received is not normally a computer file and cannot be edited by the receiver.

⊙ KEY WORD

Fax A machine capable of sending and receiving text and pictures along telephone lines.

Email

An email is an electronic message sent from one communication device (computer, telephone, mobile phone, or PDA) to another. All web browser software has email facilities. You will learn how to use the facilities offered by email in a later chapter.

Teleconferencing/videoconferencing

Videoconferencing/teleconferencing allows two or more individuals situated in different places to talk to each other in real time and see each other at the same time. They are also able to exchange electronic files with each other and pass electronic documents around rather than paper documents. It allows people to conduct "virtual" meetings. If meetings need to be conducted between people in different sites or even countries, then a WAN is used, such as the Internet.

Desktop videoconferencing systems in addition to a PC include a video compression card, a sound card, a microphone, a web cam and specialist software.

Advantages of using videoconferencing:

- Less stress as employees do not have to experience delays at airports, accidents, road works, etc.
- Improved family life, as less time spent away from home staying in hotels.
- Saves travelling time.
- Saves money as business does not have to spend money on travelling expenses, hotel rooms, meals, etc.
- Improved productivity of employees, as they are not wasting time travelling.
- Meetings can be called at very short notice without too much planning.
- More environmentally friendly as there are fewer people travelling to meetings. This cuts down on carbon dioxide emissions.
- Fewer car journeys means fewer traffic jams and hence less stress and pollution.

Disadvantages of using videoconferencing:

- The initial cost of the equipment, as specialist videoconferencing equipment is expensive.
- Poor image and sound quality.
- People can feel very self-conscious when using videoconferencing and not communicate effectively.
- Although documents and diagrams in digital form can be passed around, an actual product or component cannot be passed around.
- Lack of face-to-face contact may mean a discussion may not be as effective.
- If the delegates are in distant locations the picture can be out of synchronization with the speech which can be distracting.

Input/output devices for videoconferencing

A variety of input/output devices are used by videoconferencing systems.

- Delegates use microphones to capture their speech.
- Headphones or speakers enable delegates to listen to other delegates.
- A web cam is used to capture video images of themselves.
- A monitor/screen is used to view video images of other delegates.

Activity 4.1

Produce a poster to be put up in the computer room entitled: The advantages and disadvantages of networks

Use the Internet and information contained in the topic as your sources of information. You have to choose suitable software for this task and make it as eye-catching as you can.

Activity 4.2

Create diagrams using suitable software to produce a series of posters that show the following network topologies:

- Bus
- Star
- Tree.

If there is space, and you have time, you could add the advantages and disadvantages of each topology to your posters.

QUESTIONS C

1 Give the name of the device that is used so that several computers in a home can all share a single Internet connection. *(1 mark)*

2 a Give **two** advantages of using the Internet. *(2 marks)*
  b Give **two** disadvantages of using the Internet. *(2 marks)*

3 A company has branches all over the world and uses tele/videoconferencing to communicate with its employees.
  a Describe what is meant by tele/videoconferencing. *(2 marks)*
  b Give **two** benefits to the company and its employees of using tele/videoconferencing. *(2 marks)*

4 Explain the main difference between the network devices hub and switch. *(2 marks)*

5 A school uses a proxy server between their main server and their Internet connection.

  Explain **two** reasons why the school will connect to the Internet through the proxy server. *(2 marks)*

REVISION QUESTIONS

1 LAN and WAN are both types of computer network.
  a i What does LAN stand for? [1]
    ii What does WAN stand for? [1]
  b Give **two** differences between a LAN and a WAN. [2]
  c Give **three** advantages to computer users of a LAN, rather than working on stand-alone machines. [3]
  d Give **one** method which can be used to prevent data from being misused when it is being transferred between computers. [1]

2 When goods are ordered over the Internet, payment has to be made.
  a Give **one** method of payment used over the Internet. [1]
  b Describe why some people may not want to put their payment details into a web site. [2]
  c Describe **one** way the online store can make sure that payment details are safe. [2]

3 Explain briefly how each one of the following helps improve the security of a network.
  a User-ID [2]
  b Password [2]
  c Encryption [2]

4 A company is thinking of installing a new network. They have the choice of a wired network, where cables are used to transmit the data, or a wireless network, where no cables are needed. Describe the relative advantages and disadvantages in using a wireless network. [4]

5 Videoconferencing is used by many organizations to conduct meetings at a distance.
  a Explain **two** advantages of videoconferencing. [2]
  b Explain **two** disadvantages of videoconferencing. [2]

Test yourself

The following notes summarize this topic. The notes are incomplete because they have words missing. Using the words in the list below, copy out and complete the sentences A to J, underlining the words that you have inserted. Each word may be used more than once.

Internet	web browser	local	hackers
wireless	encrypted	Bluetooth	
wide	videoconferencing	synchronize	

A With _____ communication, the data travels through the air rather through cables.

B When making online payments, the card details are _____, which means if they are intercepted by hackers, then details will be meaningless and useless.

C _____ is a method used to transfer data over short distances from fixed and mobile devices.

D Bluetooth can be used to _____ the music on your home computer and your portable music player so the tracks stored on each are the same.

E _____area networks are those networks that are restricted to a single building or site.

F _____ area networks are situated across a wide geographical area.

G E-commerce systems use the _____ for the purchase of goods and services.

H In order to pay for online purchases, a customer must enter their credit or debit card details and many customers worry that this information could be accessed by _____.

I _____ allows virtual face-to-face meetings to be conducted without the participants being in the same room or even the same geographical area.

J _____ software is a program that allows access to web pages stored on the Internet.

EXAM AND EXAM-STYLE QUESTIONS

1 Complete each sentence below using **one** item from the list.

[4]

A hub   An intranet   A proxy server

A WAN   A WLAN

a .... is a device used to connect computers together to form a LAN.

b .... is a network with restricted access.

c .... can allow networked computers to connect to the Internet.

d .... is a wireless local area network.

(Cambridge IGCSE Information and Communication Technology 0417/11 q5 May/June 2010)

2 Aftab and his family have three computers in their home. He wants to connect the computers into a network. Explain why he would need:

A router   A browser   Email   An ISP

[4]

Cambridge IGCSE Information and Communication Technology 0417/11 q12 May/June 2010

3 A small office has four standalone computers. The office manager wants to connect the computers together to form a LAN.

a Name a network device which would have to be present in each computer before they could be networked. [1]

b Give **two** reasons why a WLAN would be preferable to a cabled LAN. [2]

c Give **two** reasons why the manager should **not** use Bluetooth technology to create the network. [2]

d The company's workers are concerned that their payroll data may not be secure as a result of the computers being networked. Explain why the workers are concerned. [6]

e Give **three** actions that the office manager could take to ensure data security. [3]

Cambridge IGCSE Information and Communication Technology 0417/13 q14 Oct/Nov 2010

4 Describe **five** differences between a WAN and a LAN. [5]

44

Data can exist in lots of different forms but ultimately the data has to be converted into digital form for the computer to process it. When we put data into a structure for processing, such as in a spreadsheet or database, we have to instruct the computer what data type we are going to use for each item of data.

The key concepts covered in this chapter are:
▸▸ Identification of different data types
▸▸ Selecting appropriate data types for a given set of data
▸▸ Understand the meaning of the terms file, record, field, and key field
▸▸ Understand the different database structures
▸▸ Understand the differences between analogue and digital data
▸▸ Understand the need for conversion between analogue and digital data

Data types

The data entered into a computer for processing is usually one of the following types:

▸▸ logical/Boolean
▸▸ alphanumeric/text
▸▸ numeric (real and integer)
▸▸ date.

Logical/Boolean

Logical/Boolean data can have only one of two values: True or False. Any data which can be stored as two possibilities such as true/false, 1 or 0, yes or no can be stored as a logical/Boolean data type.

Alphanumeric text

Alphanumeric refers to all the letters and numbers, and other characters you see on the keyboard that can be typed in such as /, @, etc. Text is just the letters on the keyboard (A to Z).

Numeric (real and integer)

Numeric numbers can be real or integer. A real number is a number which contains numbers after the decimal point (e.g., 3.45). integers are whole numbers (i.e., positive, negative whole numbers, and zero). Telephone numbers are not numeric because they often contain leading zeros (e.g. 00300....).

Date

There are lots of different ways of writing a date. The common ways include:

▸▸ dd/mm/yy, for example 12/03/11 for the date 12 March 2011
▸▸ mm/dd/yy, for example 03/12/11 for the date 12 March 2011 (as used in the US)
▸▸ yyyy-mm-dd, for example 2011-03-12

There are many other different ways you can store dates in software packages such as spreadsheets and databases.

This dialogue box allows you to change a date to a required format. Notice the part where you can set the date to the format which is used by a particular country.

QUESTIONS A

1 Data is to be stored in a structure and to do this a data type must be chosen for each item of data.
Choose a suitable data type from the following list for the data shown in the table: *(6 marks)*
 Logical/Boolean
 Alphanumeric/text
 Numeric
 Date

Name of field	Example data	Data type
Title	(Mr, Mrs, Ms, Dr, etc.)	
Phone number	0798273232	
Sex	M or F	
Country	Botswana	
Date of birth	01/10/03	
Years at address	4	

2 Here is a list of examples of data to be put into a structure.

Some of this data can have the logical/Boolean data type and some of it cannot.

Complete the following table by placing a tick in the box next to those items of data that could use a logical/Boolean data type. *(4 marks)*

Items of data	Tick if data type is logical/Boolean
Driving licence (yes or no)	
Sex (M or F)	
Size (S, M, L, XL, XXL)	
Airport code	
Car registration number	
Date of purchase	
Car type (manual or automatic)	
Fuel type (diesel or petrol)	

Putting data in a database structure

Database, spreadsheet and many other types of software require that the data is put into a structure. Once the data is put into this structure it can be manipulated and output in lots of different ways.

An organized store of data on a computer is called a database.

It is possible to create simple database structures using spreadsheet or database software. There are two types of database, called a flat-file database and a relational database. Here, you will learn about the important differences between the two types of data structure and also how data may be held more efficiently using a relational database.

Choosing the software to create a database structure

There are two types of software you could use to produce a database:

» spreadsheet software
» database software.

You can build a simple database by organizing the data in rows and columns in a table. In the database shown the columns represent each of the fields and the rows are the records.

Each column represents a field of the table.

Sex	Year	Tutor Group
M	7	Miss Hu
M	7	Mr Zade
F	8	Dr Hick
F	7	Mrs Wong
M	7	Miss Kuyt
F	8	Mr Singh

This row contains the set of the fields. Each row is a record.

Fields, records and files: what do they all mean?

There are some database terms you will need to familiarize yourselves with. These are:

File: A collection of related records is called a file. The group of records for all the pupils in the school is called the pupil file. Often a simple file holding a single set of data is called a table.

Record: All the data relating to a single thing or person is called a record. A record consists of fields.

Field: A field is a single item of data. In other words it is a fact. A surname would be an example of a field.

Data: These are facts about a specific person, place or thing.

Information: Data + Meaning = Information. For example, 12/03/11 is data. Only when we understand it is the "date for the exam" and it is in dd/mm/yy do we have the information that the exam date is 12 March 2011. So date needs "field name" and "format" to become information.

Table: In databases a table is used to store data with each row in the table being a record and the whole table being a file. When only one table is used, it is a very simple database and it is called a flat-file database.

For more complex databases, created using specialist database software, lots of tables can be used and such a database is called a relational database.

Key fields

A key field is a piece of data in a database that is unique to a particular record. For example, in a file of all the children in a school, a record would be all the data about a particular pupil. The key field would be Pupil Number, which would be a number set up so that each pupil is allocated a different number when they join the school. No two pupils would have the same number. Surname would not be unique and so is unsuitable for a key field. It is possible to have more than one key field in a record. Even pupils with the same first name and surname would have to be given a unique Pupil number.

Flat files and relational databases

Computerized databases may be divided into two types: the limited flat-file database suitable for only a few applications, and the much more comprehensive and flexible relational database.

Flat-file databases

Flat-file databases are of limited use and are only suitable for very simple databases. Flat files only contain one table of data. A record is simply the complete information about a product, employee, student, etc. This is one row in the file/table. An item of information such as surname, date of birth, product number, product name, in a record is called a field. The fields are the

	A	B	C	D	E	F	G	H	I	J	K	L	M	N	O	P	Q
1	QNo	Title	Initial	Surname	Street	Postcode	No_in_house	Type	Garden	Paper	Bottles	Cans	Shoes	Carriers	Compost	Junk_mail	
2	1	Mr	A	Ahmed	18 Rycroft Road	L12 5DR	1	S	S	Y	Y	Y	Y	Y	Y	10	
3	2	Miss	R	Lee	1 Woodend Drive	L35 8RW	4	D	M	Y	Y	Y	N	N	Y	4	
4	3	Mr	W	Johnson	42 Lawson Drive	L12 3SA	2	S	S	Y	Y	Y	N	N	Y	0	
5	4	Mrs	D	Gower	12 Coronation Street	L13 8JH	3	T	Y	Y	N	N	N	N	N	9	
6	5	Dr	E	Fodder	124 Inkerman Street	L13 5RT	5	T	Y	N	N	N	N	N	N	12	
7	6	Miss	R	Fowler	109 Pagemoss Lane	L13 4ED	3	S	S	N	N	N	N	N	N	5	
8	7	Ms	V	Green	34 Austin Close	L24 8UH	2	D	S	N	N	N	N	N	N	7	
9	8	Mr	K	Power	66 Clough Road	L35 6GH	1	T	Y	Y	Y	Y	N	N	N	7	
10	9	Mrs	M	Roth	43 Fort Avenue	L12 7YH	3	S	M	N	N	Y	N	N	N	7	
11	10	Mrs	O	Crowther	111 Elmshouse Road	L24 7FT	3	S	M	Y	Y	Y	N	N	N	8	
12	11	Mrs	O	Low	93 Aspes Road	L12 6FG	1	T	Y	Y	Y	Y	Y	N	N	11	
13	12	Mrs	P	Crowley	98 Forgate Street	L12 6TY	5	T	Y	Y	Y	Y	N	N	N	15	
14	13	Mr	J	Preston	123 Edgehill Road	L12 6TH	6	T	Y	Y	Y	N	N	N	N	2	
15	14	Mr	J	Quirk	12 Leopold Drive	L24 6ER	4	S	M	Y	Y	N	N	N	Y	2	
16	15	Mr	H	Etheridge	13 Cambridge Avenue	L12 5RE	2	S	L	Y	N	Y	N	N	Y	5	
17	16	Miss	E	James	35 Speke Hall Road	L24 5VF	2	S	L	Y	N	Y	N	N	Y	5	
18	17	Mrs	W	Jones	49 Abbeyfield Drive	L13 7FR	1	D	M	N	N	N	N	N	Y	5	
19																	

A flat file uses a single table of data set up like this.

vertical columns in a table. Because flat-file databases only contain one table, this limits their use to simple data storage and retrieval systems such as storing a list of names, addresses, phone numbers, etc. Flat files are unsuited to business applications where much more flexibility is needed. Tables consist of columns and rows organized in the following way:

» The first row contains the field names.
» The rows apart from the first row represent the records in the database.
» The columns contain the database fields.

The problems with flat-file systems

Flat files store all the data in a single table. The disadvantages of using a flat file are:

» Data redundancy. There is often a lot of duplicate data in the table. Time is wasted retyping the same data and more data is stored than needs to be, making the whole database larger.
» When a record is deleted, a lot of data that is still useful may also be deleted.

Relational databases

In a relational database, we do not store all the data in a single file or table. Instead the data is stored in several tables with links between the tables to enable the data in the separate tables to be combined together if needed.

To understand this, look at the following example:

A tool hire business hires tools such as ladders, cement mixers, scaffolding, chain saws, etc., to tradesmen. The following would need to be stored:

» data about the tools
» data about the customers
» data about the rentals.

Three tables are needed to store this data and these can be called:

» Tools
» Customers
» Rentals.

If the above were stored in a single table, (in other words using a flat file), there would be a problem. As all the details of tools, customers and rentals are stored together there would be no record of a tool unless it had been hired by a customer. There would be no record of a customer unless they had hired a tool at the time.

In the flat file there would be data redundancy because customer address details are stored many times for each time they hire a tool. This means the same data appears more than once in the one table. Hence there are serious limitations in using flat files and this is why data is best stored in a relational database where the data is held in several tables with links between the tables.

⊙ KEY WORDS

Data redundancy where the same data is stored more than once in a table or where the same data is stored in more than one table.
Key field this is a field that is unique for a particular record in a database.

Creating a database

Before you create a structure for a database it is important to look at a sample of the data that needs to be stored.

A school keeps details of all its pupils in a database. As well as personal details (name, address, etc.), the school also holds details of the tutor group and tutors.

The person who is developing the database asks the principal for a sample of the data. This sample of the data is shown below:

Description of data stored	Sample data
Pupil number	76434
Surname	Harris
Forename	Amy
Date of birth	15/03/98
Street	323 Leeward Road
Town	Waterloo
Postcode	L22 3PP
Contact phone number	0151-002-8899
Home phone number	0151-002-1410
Tutor Group	7G
Tutor number	112
Tutor teacher title	Mr
Tutor surname	Harrison
Tutor initial	K

Three tables are used with the names: Pupils, Tutor Groups and Tutors.

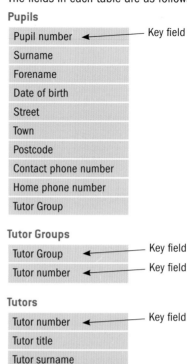

The fields in each table are as follows:

Pupils

Pupil number — Key field
Surname
Forename
Date of birth
Street
Town
Postcode
Contact phone number
Home phone number
Tutor Group

Tutor Groups

Tutor Group — Key field
Tutor number — Key field

Tutors

Tutor number — Key field
Tutor title
Tutor surname
Tutor initial

The data is put into three tables rather than one because it saves time having to type the same details over and over about the tutor for each pupil. In other words it reduces data redundancy. If there are 25 pupils in each form, the tutor's details (i.e., Tutor number, Tutor title, Tutor surname, etc.) would need to be entered 25 times. If instead we put these details in their own table, we can access them from the Tutor Group field and we only need to type in the Tutor Group details once.

⊙ KEY WORDS

Relational database a database where the data is held in two or more tables with relationships (links) established between them. The software is used to set up and hold the data as well as to extract and manipulate the stored data.
Relationship the way tables are related or linked to each other.

QUESTIONS B

1 A luxury car rental firm keeps the details of the cars it rents out in a table. The structure and contents of this table are shown below.

Reg-number	Make	Model	Year
DB51 AML	Aston Martin	DB7	2009
CAB 360M	Ferrari	360 Modena	2008
GT X34 FER	Ferrari	355 Spider	2000
MAS 12	Maserati	3200 GTA	2001
FG09 FRT	Porsche	911 Turbo	2009
M3 MMM	BMW	M3 Conv	2010
T433 YTH	Jaguar	XK8	2009

a Give the names of **two** fields shown in the above table. *(2 marks)*
b Give the name of the field that should be chosen as the key field. *(1 mark)*
c Explain why the field you have chosen for your answer to part (b) should be chosen as the key field. *(1 mark)*
d How many records are there in the table? *(1 mark)*

➔

2 Most schools now use databases to store details about each pupil. The table shows some of the field names and data types stored in one pupil database.

Field name	Data Type
UniquePupilNumber	Number
Firstname	
Surname	Text/Alphanumeric
FirstLineAddress	Text/Alphanumeric
SecondLineAddress	Text/Alphanumeric
Postcode	
LandlineNo	Text/Alphanumeric
DateOfBirth	Date
FreeSchoolMeals(Y/N)	

a Give the most appropriate data types for the following fields: *(3 marks)*
   i Firstname
   ii Postcode
   iii FreeSchoolMeals (Y/N)
b Give the names of **three** other fields that are likely to be used in this database. *(3 marks)*

c Explain which field is used as the key field in the database and why such a field is necessary. *(2 marks)*
d It is important that the data contained in this database is accurate.
   Describe how **two** different errors could occur when data is entered into this database. *(2 marks)*
e Explain how the errors you have mentioned in part (d) could be detected or prevented. *(2 marks)*

3 a Explain what is meant by a flat-file database. *(1 mark)*
b Explain what is meant by a relational database. *(1 mark)*
c Describe an application where a flat-file database would be suitable. *(2 marks)*

4 Databases are of two types: flat file and relational.
a Describe **two** differences between a flat-file database and a relational database. *(2 marks)*
b A dress hire company needs to store details of dresses, customers, and rentals. They want to store these details in a database.
   Which type of database do you suggest and give **two** reasons for your answer. *(3 marks)*

Linking files or tables (i.e., forming relationships)

To link two tables together there needs to be the same field in each table. For example, to link the Pupils table to the Tutor Groups table we can use the Tutor Group field as it is in both tables. Similarly, the Tutor Groups table and the Tutors table can be linked through the Tutor number field. Links between tables are often called relationships, and they are one of the main features of relational databases.

Pupils

Pupil number
Surname
Forename
Date of birth
Street
Town
Postcode
Contact phone number
Home phone number
Tutor Group

Tutor Groups

Tutor Group
Tutor number

Tutors

Tutor number
Tutor title
Tutor surname
Tutor initial

Foreign keys

A foreign key is a field of one table which is also the key field of another. Foreign keys are used to establish relationships between tables. In the above example the field Tutor Group would be the key field in the Tutor Groups table and a foreign key in the Pupils table.

The advantages of relational databases

Using a relational database, means that you do not have to type in all the data for each pupil when you create the Tutor Groups table.

This has the following advantages:

▸▸ It saves time typing.
▸▸ It reduces typing errors.
▸▸ It therefore reduces redundancy.

Analogue to digital conversion

Analogue quantities are continuously variable, which means that they do not jump in steps from one value to another. Temperature is an analogue quantity because temperatures do not jump from one degree to the next as there are many values in between. Other examples would be height, weight, light level and humidity.

Analogue to digital converters (ADC)

Digital quantities jump from one value to the next. Computers are nearly always digital devices and can only operate with digital values. A quantity such as temperature can vary over a whole range of values, for example they could be 5, 5.1, 5.01, 5.001, and so on. As you can see, analogue quantities can have an infinite range of values. If analogue values such as temperature readings need to be input into a computer, they first need to be converted to digital values using an analogue to digital converter (ADC).

Digital to analogue converters (DAC)

If the output signals from a computer are used to control a device, such as an opening vent in a greenhouse, they will need to be converted from digital to analogue. This is performed using a device called a digital to analogue converter (DAC).

The need for conversion between analogue and digital data for sound signals

Sound is a continuous wave and is therefore an analogue signal. As computers can only work with digital data (i.e., data stored as numbers), it is necessary to convert the analogue data to digital data using an analogue to digital converter. Once the sound data is in digital format, the computer can store it, edit it, transfer it to another computer and play it back. Loudspeakers and headphones use analogue signals, so when these devices are used to output sound from the computer, the digital data/signal from the computer has to be changed to an analogue signal using a digital to analogue converter.

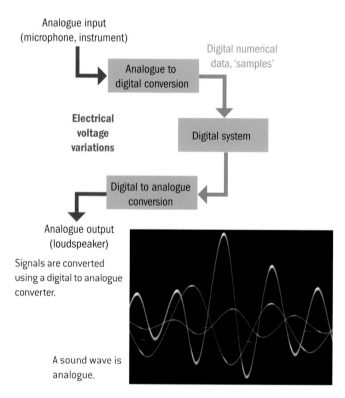

Analogue input
(microphone, instrument)

Digital numerical
data, 'samples'

Analogue to
digital conversion

**Electrical
voltage
variations**

Digital system

Digital to analogue
conversion

Analogue output
(loudspeaker)

Signals are converted using a digital to analogue converter.

A sound wave is analogue.

Temperature control in a greenhouse

Temperature is an analogue quantity and an ADC will be needed so that the temperature signal can be converted to a digital signal before it can be processed by a computer. A comparison with the desired temperature reading can send an output signal to a heater to turn it either on or off. Temperature control systems are usually more complicated than this because in the summer, the temperature in the greenhouse could go high unless there is some way of cooling it down. Usually this is done by the computer controlling vents (like windows) or by using an air-conditioning system.

Test yourself

The following notes summarize this topic. The notes are incomplete because they have words missing. Using the words in the list below, copy out and complete the sentences A to N, underlining the words that you have inserted. Each word may be used more than once.

> analogue to digital   record   digital to analogue
> foreign   real   key field   Boolean   links
> redundancy   fields   relational   analogue
> alphanumeric   digital   table

A Numeric data can be in two forms integer or _____.

B Letters and numbers together are called _____ data.

C Logical or _____ data can have only one of two values: True or False.

D A _____ is a piece of data in a database system that is unique to a particular record.

E All the data relating to a single thing or person is called a _____.

F A record consists of many _____.

G A database where the data is held in two or more tables with links established between them is called a _____ database.

H Flat-file databases only contain one _____ of data.

I Relational databases have _____ called relationships between the tables.

J In order to create a relationship a primary key in one table is linked to the same field called a _____ key in the other table.

K Flat files may suffer from data _____ where a lot of the data in the table is repeated.

L An analogue signal from a temperature sensor will need to be converted into a _____ signal before it can be processed by a computer.

M Changing from an analogue to a digital signal is performed using an _____ converter.

N In order to control a device such as a motor, a digital signal from the computer will need to be changed into an _____ one and this is done using a _____ converter.

REVISION QUESTIONS

1 Here is a sample of the data that is to be stored in an employee database. The data items shown are the employees' surname, initial, street, postcode, and telephone number.

Ahmed, V, 123 The High Street, L23 6DE, 0151-002-1112

Delos, N, 64 North Way, L9 8SS, 0151-002-0011

Doyle, B, 12 Crosby Road, L23 2DF, 0151-002-1212

Carrol, A, 15 Barkfield Drive, L23 7YH, 0151-002-0899

Conway, T, 6 Windle Hey, L23 6ER, 0151-003-0899

Hoch, J, 4 Empress Road, L22 7ED, 0151-003-9090

Hoch, J, 4 Empress Road, L22 7ED, 0151-003-9090

a A table is to be set up with four fields. Give field names for the **four** fields that would be suitable for the above set of data items. [1]

b The person who is designing the database looks at the sample of data above and notices that there are two people with the same surname and initial who live at the same address.

   i Explain why the surname would be an unsuitable key field. [1]

   ii It is decided that each employee should be given a unique number. What would be a suitable field name for this field? [1]

2 A store keeps details about their customers and their orders in an information handling system. Part of the data in this system is shown below:

Customer number	Name	Item code	Size	Cost	Delivery
2314	J.Hughes	464	Small	$290	N
9819	D. Wong	255	Large	$767	Y
1311	C. Khled	747	Small	$239	N
8276	K. Lee	299	Small	$200	Y
9223	F. Smith	108	Large	$823	Y

a Give the names of **two** key fields used in the above table. [2]

b There is **one** "Boolean" field in this database. Give the name of it. [1]

c Give the names of **two** other fields that could be sensibly used in this table other than customer name, customer address, and customer telephone number. [2]

d The store offers free delivery on all items with a cost greater than $200. How many of the customers shown would qualify for the free delivery? [1]

3 Sensors such as temperature, moisture, and light sensors are used to monitor the growing conditions in a greenhouse.

a Explain why computers are not able to read the data directly from these sensors. [2]

b Give the name of the device needed to enable the computer to read this data. [1]

EXAMINATION QUESTIONS

1 A school uses a computer-controlled greenhouse to grow plants.

  a Name **three** sensors that would be used in the greenhouse. [3]

  b Explain why analogue to digital conversion is needed when computers are used to control a greenhouse. [2]

Cambridge IGCSE Information and Communication Technology 0417/11 q10 May/June 2010

2 A school library has a file for storing details of the books it has and a file for storing details of its borrowers. The two files are linked using a common field.

Book File

Code	Title	Author	Published	Number in stock	Cost
1857028898	The Code Book	Simon Singh	1999	2	£10.99
0747591054	The Deathly Hallows	J K Rowling	2007	8	£17.99
0748791167	Sepulchre	Kate Mosse	2007	3	£18.99
0563371218	Full Circle	Michael Palin	1997	1	£19.99

Years 10 and 11 Borrower File

Number	Name	Form	Book borrowed	Due back
0102	Me Te Loan	11A	1857028898	14/06/2012
1097	Gurvinder Moore	10C	0747591054	12/06/2012
0767	Akhtar Aftab	10B	0748791167	06/06/2012
0611	Graham Reeves	10D	0563371218	08/06/2012

  a How many records are there in the book file? [1]

  b How many fields are there in the borrower file? [1]

  c What type of database do these two files form? [1]

  d Give **two** reasons why this type of database system is used rather than having two flat files. [2]

  e Which field is the primary key in the Book file? [1]

  f Which field is the foreign key? [1]

  g Which field has the data type currency? [1]

  h What data type would be most appropriate for the **Due Back** field? [1]

Cambridge IGCSE Information and Communication Technology 0417/11 q11 May/June 2010

3 A geography teacher wants to use a computer to monitor the weather conditions using the school weather station.

  a Name **three** sensors in the weather station. [3]

  b Explain why computers are unable to read the data directly from these sensors. [2]

  c What device is needed to enable the computer to read the data? [1]

(Cambridge IGCSE Information and Communication Technology 0417/13 q19 Oct/Nov 2010)

The effects of using ICT

In this chapter you will learn about some of the many impacts that the use of ICT has on our lives. Most of these impacts are for the good but there are some bad ones too. You will be looking at the way ICT has changed employment and the way that we work and you will look at the impact that using the Internet has had on people being able to work from home. You will also be looking at videoconferencing and how it is used and also some of the health and safety issues when working with ICT equipment.

The key concepts covered in this chapter are:
▶▶ Negative aspects in using ICT such as hacking, viruses, and illegal copying of software
▶▶ The effect of ICT on patterns of employment
▶▶ The effects of microprocessor-controlled devices in the home
▶▶ The capabilities and limitations of ICT
▶▶ The use of Internet developments (Web 2.0, blogs, Wikis, etc.)
▶▶ Issues concerning material on the Internet
▶▶ Potential health and safety problems

Negative aspects in using ICT

There is no doubt that the use of ICT has brought many benefits to society. Unfortunately the use of ICT has also brought a number of problems and here we will look at some of them.

Software copyright theft

People and organizations spend time and money developing new software so they have a right to have their work protected from being copied and used by others without their permission. This is called software copyright. In most countries there are laws that help prevent software owners from having their work copied or used by others illegally (e.g. without payment).

Software copyright covers the following:

▶▶ Software being copied and sold or given to others.
▶▶ Using the name of copyrighted software if you are not permitted to do this.
▶▶ Using software on a network of more users that you have paid for (e.g. 40 users on a 20-user licence).
▶▶ Software being copied and then amended and passing it off as your own.

Computer viruses

Viruses pose a major threat to ICT systems. A virus is a program that replicates (i.e., copies) itself automatically and can cause harm by copying files, deleting files or corrupting files. Once a computer or media has a virus copied onto it, it is said to be infected. Most viruses are designed to do something apart from copying themselves. For example, they can:

▶▶ display annoying messages on the screen
▶▶ delete programs or data
▶▶ use up resources, making your computer run more slowly
▶▶ spy on your online use – for example, they can collect usernames and passwords, and card numbers used to make online purchases.

One of the main problems is that viruses are being created all the time and that when you get a virus infection, it is not always clear what a new virus will do to the ICT system. Apart from the damage many viruses cause, one of the problems with viruses is the amount of time that needs to be spent sorting out the problems that they create. All computers should be equipped with anti-virus software, which is able to detect and delete these viruses.

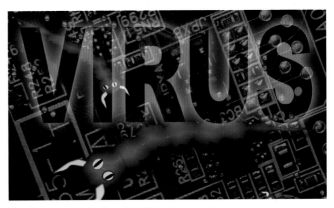

Viruses pose a major threat to all computer systems.

Preventing a virus attack
To prevent a virus attack you should:

▶▶ install anti-virus software
▶▶ not open file attachments to emails unless you know who they are from
▶▶ not allow anyone to attach portable drives or memory sticks to your computer unless they are scanned for viruses first
▶▶ not download games and other software from a site on the Internet unless it is a trusted site.

Virus scanning software should be kept up-to-date and scans should be scheduled so that they are performed automatically on a regular basis.

Viruses checking software should be set to update automatically.

Hacking and hackers

Hacking is the process of accessing a computer system without permission. Hackers often access a computer system using the Internet. In most cases the hacker will try to guess the user's password or obtain the password another way.

Some hackers just hack to see if they can enter a secure system while others do it to commit fraud by making alterations to the data, etc. In most countries hacking is a crime but it is quite hard to prove as it is necessary for it to be proved that the hacker deliberately hacked into the system.

Sometimes special software is installed on a user's computer without their permission and it can log the keys that the user presses, which can tell a hacker about usernames, passwords, emails, etc. This software is called spyware or key-logging software.

Firewalls protect networks from hackers.

Protecting against hackers

Use a firewall – a firewall is hardware, software or both that will look at the data being sent or being received by your computer to either allow it through or block it.

The main purposes of firewalls are:

▸▸ To prevent hackers accessing a computer or network using the Internet.
▸▸ To prevent undesirable content (e.g., pornography, violent videos, etc.) being accessed by users using the Internet.

"The identity I stole was a fake!
Boy, you just can't trust people these days!"

Protecting access by using user-IDs and passwords

System security is making sure that only authorized people access a computer system. This is normally done using a system of user identification (i.e., user-IDs) and passwords. System security is provided by computer software. Most computer operating systems provide systems of user-IDs and passwords.

Identifying the user to the system: user-IDs

A user-ID is a name or number that is used to identify a certain user to the network. The person who looks after the network will use this to allocate space on the network for the user. It is also used by the network to give the user access to certain files.

The network manager can also keep track of what files the user is using for security reasons. Network managers can also track improper use of the Internet using user-IDs.

Preventing unauthorized access to the system: the use of passwords

A password is a string of characters (letters, numbers, and punctuation marks) that the user can select. Only the user will know what the password is. When the user enters the password, it will not be shown on the screen. Usually an asterisk is shown when you enter a character. Only on entry of the correct password will the user be allowed access to the network. Some networks are set up so that if someone attempts to put in a password and gets it wrong three times, the user account is locked.

Encryption

Data is often encrypted (i.e. put into a secret code) before being sent over Internet or when stored on a computer. If intercepted or accessed by unauthorized people, the data cannot be understood. Credit card details and other financial details are nearly always encrypted before being sent using the Internet.

Advice about passwords

Passwords should:

▸▸ be changed regularly

▸▸ not be obvious such as your name, your dog, your favourite singer, etc.

▸▸ not be written down (e.g., on a bit of paper in your drawer, in your diary or on a note stuck on the screen).

QUESTIONS A

1 A network manager is worried about hacking and computer viruses.

    a What is meant by a computer virus? *(2 marks)*

    b What is meant by hacking? *(2 marks)*

2 Computer viruses pose a serious threat to all computer systems.

    a Give the name of the piece of software that can be used to check for viruses and also remove them.
    (1 mark)

    b Describe **two** actions that may be taken, other than using software, to prevent viruses from entering a computer system. *(2 marks)*

3 A password normally has to be entered in order to gain access to a network.

    a Give **one** reason why a password is needed. *(1 mark)*

    b Give **one** reason why this password should be changed regularly. *(1 mark)*

The effects of ICT on patterns of employment

Many businesses now operate 24 hours per day and 7 days a week, so people need to be more flexible in their hours of work. They will no longer spend their whole life doing the same job, so constant retraining will be needed so that they can take advantage of new ICT developments. Many organizations operate internationally and some jobs may be transferred abroad if the wage costs are lower in another country. Location is no longer an issue with ICT as data can be accessed from anywhere in the world.

Here are the main effects of ICT on employment:

▸▸ Fewer people needed to complete the same amount of work.

▸▸ Increased number of people working from home (called telecommuting or teleworking) using ICT equipment.

▸▸ More automation in factories due to the introduction of robots for welding, paint spraying, and packing goods.

▸▸ Continual need for training as ICT systems change.

▸▸ More availability of part-time work as many organizations need to be staffed 24/7.

▸▸ More variation in the tasks undertaken and staff need to be flexible and well trained to cope with this.

▸▸ Fewer "real" meetings as "virtual" meetings using videoconferencing are used to reduce travel time and travel costs.

▸▸ Increase in the number of technical staff needed such as network managers.

Areas of work where there is increased unemployment

There are many types of jobs where the numbers of people employed have been reduced over the years and some jobs which no longer exist. Here is a summary of these:

▸▸ Many manual repetitive jobs such as paint spraying, welding, packing goods, assembly work in factories have been replaced by robots.

▸▸ Numbers of people who work in shops have reduced because of an increased number of people shopping online. Online stores are heavily automated (e.g. many use robots to select goods and pack them) and employ fewer staff.

▸▸ Call centre work. Many jobs in call centres have been transferred to other countries where wage costs are lower.

▸▸ Designing and producing CDs/DVDs. Music, games, and software are more likely to be downloaded rather than bought on physical media such as CD. This eliminates the need for packaging.

These robots have replaced people in welding panels together in a car factory.

Areas of work where ICT has increased employment

New jobs that have been created through ICT include:

- Network managers/administrators – these are the people who keep the networks running for all the users and see to the taking of backup copies.
- Web site designers – these are the people who design and create web sites for others, as well as keep them up-to-date by adding new and deleting old material.
- Development staff – these include systems analysts (who design new ICT systems) and programmers who write the step-by-step instructions (i.e., the programs) that instruct the computers what to do.

The effects of microprocessor-controlled devices in the home

Microprocessors are computers, usually on a single chip, that are put into electronic devices to check, regulate and control something. If you think of any device in the home that needs to be controlled in some way, then it is likely to contain a microprocessor for this control. Here are some of the many devices used in the home that contain microprocessors:

- Computer systems contain one or more microprocessors for the main processing and this is called the CPU. There are also other microprocessors found in devices such as modems, routers, disk drives, etc.
- Washing machines – used to control valves to let the water in, motors to turn the drum, pumps to pump the water out, heaters to heat water up, and so on.
- Children's toys – use microprocessors to control lights, motors, speakers, etc.
- Heating systems – control the time the heating comes on, keeps the temperature constant by turning the heating on/off, etc. Modern heating systems can control the individual temperatures of each room from a central place.
- Alarm systems – can detect the presence of a burglar and some systems will even contact the police automatically.
- Digital cameras – to control the shutter speed, aperture, etc. The use of a microprocessor makes it much easier to take a good picture.
- Intelligent ovens and microwaves – some of these incorporate bar code readers which read the barcodes on packaging which the oven uses to find out the ideal time and temperature for cooking.
- Intelligent fridges – a large Internet food supply company has developed a fridge that can scan the "use by" dates on food and then move any food product to the front of the fridge automatically using a series of computer controlled panels. This reduces food wastage.

The effects of microprocessor-controlled devices on leisure time

Many microprocessor-controlled devices save us time and allow us to have more leisure time. For example, dishwashers can wash the dishes for us while we go out or watch a game of football.

Here are some ways in which microprocessor-controlled devices affect leisure time:

- Mobile phones, laptops, netbooks, PDAs, etc., mean employees can do more work on the move and at home, which can mean that they have less time for family life.
- Electrical appliances controlled by microprocessors can complete tasks such as washing and drying clothes, washing dishes, and cooking meals, which can leave a person free to do other things.
- People can become lazy and rely on machines.
- Lack of fitness as people are not spending time doing manual work.
- Can use the extra time saved to go to the gym or exercise more.
- Many people spend their leisure time playing games, surfing the net, downloading and listening to music, etc., on computers and games consoles, which all make use of microprocessors.
- Watching programmes on TV using satellite or cable TV makes use of microprocessors.
- You don't have to be in when washing is being done so you spend time at work or doing other things.

The effects of microprocessors on social interactions

The use of microprocessor-controlled ICT equipment such as computers and mobile phones has led to a huge increase in the number of ways we can keep in touch with friends and family. Social interaction is important for a person's well-being and there are a number of ways that microprocessor-controlled devices help and here are some of them:

- Playing many computer games or chatting to people in chat rooms may be to the detriment of interacting with real people. Social skills can be affected by too many solitary activities.
- Mobile phones, SMS (texting), email, blogs, social networking sites, chat rooms, and instant messaging make it much easier to keep in touch with friends and family.
- It is easy to make new friends in chat rooms and using social networking sites.

▸ If you have an unusual hobby or interest then it is much easier to find others with the same interests using the Internet.

▸ Cheap Internet phone calls, made using a service called VoIP, enable people to keep in contact with friends and family who may live in other countries.

▸ Disabled or elderly people who are confined to their home can make friends with others.

▸ It is much easier to find out about social activities, e.g. classes, concerts, activities, etc., using the Internet.

▸ Time spent not having to do jobs manually enables the time to be used to do other things.

There are a number of disadvantages and these include:

▸ People rely on the Internet for performing their daily tasks (e.g., shopping, banking, entertainment, etc.) and this means they do not meet people, which can lead to social isolation.

▸ Playing computer games is unhealthy compared to playing proper physical games and can be isolating.

▸ People can lose some basic housekeeping skills so when the machine breaks down they do not know what to do.

The effect of microprocessors on the need to leave the home

There are many ways in which microprocessor-controlled devices help people perform many tasks from home. This may be a convenience for many but it can mean the difference between having or not having your independence if you are disabled and housebound.

Here are some of the ways the use of microprocessor-controlled devices reduces the need to leave the home:

▸ Online shopping – means you order goods using the Internet and have them delivered to your home.

▸ Downloads – items such as software, music, games, ring tones, videos can be downloaded onto your computer using the Internet. There is no need to go to the shops for these.

▸ Online banking – you can view statements, transfer money between accounts, pay bills, apply for loans, etc., all from the comfort of your home.

▸ Research – many people need to find out information in their daily lives such as train times, opening times of shops, reviews of products, etc. All this can be done from home using the Internet.

▸ Entertainment – there are so many ways to entertain yourself using social networking sites, playing games, listening to music, watching videos, etc. All of these are possible using microprocessor-controlled devices.

▸ Working from home (called teleworking or telecommuting) by making use of computers and telecommunication devices and systems.

▸ No need to be present to supervise devices like ovens or washing machines.

▸ Can set TV recording or set ovens from a remote place using embedded technology and mobile phones.

QUESTIONS B

1 ICT has replaced or changed many jobs.
   a Give the names of **two** types of job that have been replaced by ICT. *(2 marks)*
   b Some jobs have changed their nature due to the introduction of ICT. Name **two** jobs where this has happened. *(2 marks)*
   c The increase in high-speed broadband links has led to cheaper international telephone calls using the Internet. Name **one** job that may be lost to a different country as a result of this. *(1 mark)*

2 The widespread use of ICT has had a huge impact on society. One benefit that it has brought is the creation of new and interesting jobs.
   a Give **three** examples of jobs that have been created through the introduction of ICT. *(3 marks)*
   b Many people have had to be retrained to cope with the introduction of new ICT systems. Explain why regular retraining is needed in the workplace. *(2 marks)*

The capabilities and limitations of ICT

ICT systems have many capabilities which are summarized here:

▸ Speed – computers are extremely fast at performing calculations and this is important where a huge amount of data is used to produce information. An example of this would be the use of millions of pieces of data from satellites and weather stations used to produce accurate weather forecasts.

▸ Accuracy – computers, when provided with accurate data, can produce accurate answers provided they have been correctly programmed.

▸ Processing power – many processors are able to do several tasks at the same time.

▸ Ability to network computers – a whole new range of services such as email, the Internet, file transfer, etc., is available when computers are linked together.

▸ Huge amount of data storage – you can store huge databases of information in the form of text, images, music, video, etc.

▸ Ability to perform searches – if a huge amount of information is stored then there needs to be a way of quickly accessing this information. Software such as web browser software or database software can find stored information quickly.

ICT systems do have a number of limitations such as the following:

▸ Bandwidth – this represents the speed with which data can be transferred over the Internet. Bandwidth limits the speed of access to the Internet.

▸ Battery life – many portable devices are limited by their battery life. Ideally netbooks, laptops, PDAs, etc., need a long battery life between charging.

- ▸ Weight – many portable devices are heavy, which limits their use. The use of flexible screens and the use of the Internet to store programs and data will mean portable devices can be lighter.
- ▸ User interfaces – users need to interact with devices such as computers and PDAs. Current methods involve using keyboards, mice, touch screens, etc. User interfaces need to be made easier and operate in a similar way to the way humans interact with each other.

Internet developments

The Internet is constantly evolving and **Web 2.0** refers to the latest developments to the Internet which include the following:

- ▸ blogs
- ▸ wikis
- ▸ digital media uploading web sites
- ▸ social networking web sites.

Blogs/web logs – are online diaries of events or journals. Blogs can be about anything. Groups, singers, and celebrities have blogs which let people know about their life and what they are doing. Blogs are also used by politicians, and for collecting public opinion about certain topics.

Wikis – a wiki is a web page that can be viewed and modified by anyone who has a web browser. This means if you have web browser software then you are able to change the content of a web page. The problem with this is that you can alter the web page to post offensive messages, etc. You will have seen the online encyclopaedia Wikipedia. This encyclopaedia has been created by ordinary people and anyone can add material or delete material from it. You may think this is bad idea, but if someone posts incorrect information then there are plenty of people around who will view it and correct it. This means that information from sites like Wikipedia may contain inaccurate information – you should always check the information with other more reliable sources.

Digital media uploading web sites – these web sites allow anyone to upload their videos, digital images, etc., to a web site which allows others to access the media.

The main advantage of these sites is that you do not have to take up storage on your own computer and it is easy to direct people to the site if they want to view your photographs. Examples of such sites include YouTube for video clips and Flickr for photographs.

Social networking web sites – these are virtual communities of people who communicate about a particular subject or interest or just make friends with each other. Members of these sites create their own profiles with information about themselves such as hobbies, likes, and dislikes, etc. You have to ensure that you do not reveal too much information about yourself as others view this information. Social networking web sites enable members to communicate using instant messaging, email, a type of blog, and even voice or videoconferencing. The services allow members to invite members and non-members into their circle of friends.

Social networking web sites you might be familiar with include:

- ▸ Facebook
- ▸ MySpace
- ▸ Twitter.

Social networking sites allow you to increase your circle of friends.

Issues with information found on the Internet

Just because information is on the Internet does not make it true. Any information you get off the Internet needs to be treated with caution. In this section you will be looking at how you might assess the worth of the information and how you can check its accuracy.

Unreliability of information on the Internet

Many people think that because information is published on the Internet it must be true. This is far from the truth, as there are many examples of sites that contain material which is completely untrue. Some sites deliberately set out to misinform or deceive.

It is important to remember that anyone is able to produce a web site and publish it on the Internet. They do not have to be an authority on the subject the site is about. In some cases the person creating the site has not checked the information on the site.

Advantages of using information on the Internet are:

- ▸ The information is up-to-date compared to many books.
- ▸ It is quick to search for the information.

>> It is available on your home computer or mobile device.
>> You can obtain information from other parts of the world easily.

Disadvantages of using information on the Internet are:

>> The information could be unreliable
>> There could be too much information to read (e.g. too many hits).

How can you make sure that information you use is accurate? Here are a few steps you can take:

>> Check the date that the site was last updated. Bogus sites are not updated very often. Also sites go out-of-date, so you need to be sure that the site you use has been updated recently.
>> Only use sites produced by organizations you have heard of (e.g., newspapers, BBC, etc.).
>> Use several sites to get the information and check that the sites are giving the similar information.
>> Follow the links to see if they work. Many bogus sites have links that do not work.

Activity 6.1

Looking at bogus sites
Here are some bogus sites. Just a bit of fun really, but remember than some bogus sites are there to steal your identity or your money and others may give people wrong information.

1   http://home.inreach.com/kumbach/velcro.html
2   http://www.weathergraphics.com/tim/fisher/
3   http://www.geocities.com/Heartland/Acres/3072/camera1.html
4   http://www.genochoice.com/
5   http://www.dhmo.org/

Undesirability of information on the Internet

It is impossible to censor what is on the Internet. This is because much of the information on the Internet comes from other countries. Unfortunately the Internet contains pornography, violent videos, sites that promote racial hatred, and so on. Schools and parents are able to use parental controls that restrict the sites and information that can be viewed on the Internet.

The security of data transfer

Once networks or individual computers are connected to the Internet there is a chance that others may try to hack into the system usually in order to commit fraud. In this part you will be looking at some of the issues concerned with the security of data transfer.

Phishing

Phishing is fraudulently trying to get people to reveal usernames, passwords, credit card details, account numbers, etc., by pretending you are a bank, building society, or credit card company, etc. Emails are sent to you pretending to be from a bank, building

society, credit card company, etc. They usually say that there has been a problem with your account and ask you to update (reveal) information such as passwords, account details, etc. Under no circumstances should you reveal this information. If you do then these details will be used to commit fraud.

Pharming

Pharming is where malicious programming code is stored on a computer. Any users who try to access a web site which has been stored on the computer will be re-directed automatically by the malicious code to a bogus web site and not the web site they wanted. The web site they are directed to looks almost the same as the genuine web site. In many cases the fake or bogus web site is used to obtain passwords or banking details so that these can be used fraudulently.

Spam

Spam is email that is sent automatically to multiple recipients. These emails are unasked for and in the main about things you are not interested in. Spam is also called junk email and in most countries, the sending of spam is not illegal despite organizations wasting huge amounts of time getting rid of it.

There are a number of problems with spam:

>> It can take time looking at the email names of spam before deleting it.
>> Spam can sometimes be used to distribute viruses.

People who collect email addresses for the purpose of sending spam are called spammers. They collect email addresses from chat rooms, mailing lists, social networking sites, etc.

Spam filters are software that is used to filter out spam from legitimate email. The problem is that some spam gets through and some legitimate email ends up as spam. This means you cannot just delete all the emails in the spam folder in one go.

> ## ⊙ KEY WORDS
>
> **Phishing** tricking people into revealing their banking or credit card details using an email.
> **Spam** unsolicited bulk email (i.e., email from people you do not know, sent to everyone in the hope that a small percentage may purchase the goods or services on offer).

Spam is trapped by the spam filter and put into its own folder.

The potential health hazards when using computers

As there are potential hazards when using computers and other ICT equipment, you need to be aware of what the hazards are. You also need to be aware of the symptoms of the medical conditions they can cause.

The main health problems are:

- Repetitive strain injury (RSI) – this is caused by typing or using a mouse over a long period of time. RSI is a painful illness that causes swelling of the wrist or finger joints. It can get so bad that many sufferers are unable to use their hands.
- Eye strain – looking at the screen all day can give you eye strain. Many of the people who use computer screens for long periods have to wear glasses or contact lenses. The symptom of eye strain is blurred vision.
- Headaches caused by glare on the screen, a dirty screen or screen flicker.
- Back and neck ache – is a painful condition that prevents you from sleeping properly and doing many activities such as playing sport.

Back ache is a common ailment in computer users.

Methods of preventing or reducing the risks of health problems

Back and neck ache

The following can cause back and neck ache:

- Not sitting up straight in your chair (i.e., incorrect posture).
- Using a laptop on your knee for long periods.
- Working in cramped conditions.

To help prevent back and neck problems:

- Use an adjustable chair (NB in workplaces in some countries this is a legal requirement but you should ensure that the chair you use at home is adjustable).
- Always check the adjustment of the chair to make sure it is suitable for your height.
- Use a foot support, called a footrest, if necessary.
- Sit up straight on the chair with your feet flat on the floor or use a foot rest.
- Make sure the screen is lined up and tilted at an appropriate angle so you are looking directly across and slightly down – never to the side or up.

© 1998 Randy Glasbergen.

"Suspending your keyboard from the ceiling forces you to sit up straight, thus reducing fatigue."

Repetitive strain injury (RSI)

The following can cause RSI:

- Typing at a computer for a long time.
- Using a mouse for long periods.
- Not adopting correct posture for use of mouse and keyboard.
- Not having properly arranged equipment (e.g., keyboard, mouse, screen, etc.).

To help prevent RSI:

- Take regular breaks.
- Use an ergonomic keyboard.
- Make sure there is enough space to work comfortably.
- Use a document holder.
- Use an ergonomic keyboard/mouse.
- Use a wrist rest.

- Keep your wrists straight when keying in.
- Position the mouse so that it can be used keeping the wrist straight.
- Learn how to type properly – two finger typing has been found to be much worse for RSI.

Eye strain/headaches

The following can cause eye strain/headaches:

- Using the screen for long periods.
- Working without the best lighting conditions.
- Glare on the screen/screen flicker.
- Dirt on the screen.

To help avoid eye strain/headaches:

- Take regular breaks (e.g. every hour)
- Keep the screen clean, so it is easy to see characters on the screen.
- Use appropriate lighting (fluorescent tubes with diffusers).
- Use blinds to avoid glare.
- Give your eyes a rest by focusing on distant objects.
- Have regular eye-tests (NB if you use a screen in your work, then your employer may be required by law in some countries to pay for regular eye-tests, and glasses if they are needed).
- Ensure that the screen does not flicker.
- Change to an LCD screen if you have an old CRT monitor.

Safety issues and measures for preventing accidents

There are a number of safety issues related to using computers and these include the following:

- Overheating
  Causes – computers give out large amounts of heat and rooms containing them can become unbearably hot in the summer.
  Prevention – install air-conditioning.

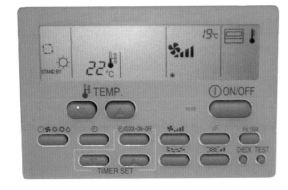

Air conditioning allows you to adjust the temperature in a room.

- Fire
  Cause – computers need lots of power sockets. ICT equipment uses a lot of power sockets and if multi-sockets are used then it is easy to overload the mains circuit. This is dangerous and could cause a fire.
  Prevention – computer rooms should be wired specially with plenty of sockets. Fire extinguishers should be provided and many large organizations use sprinkler systems that activate automatically. Smoke detectors should also be used.

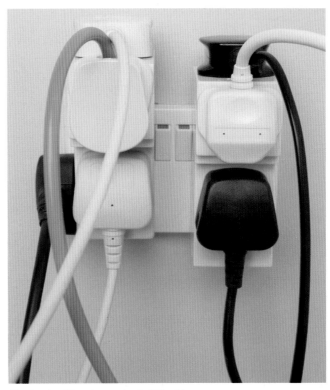

Circuits should not be overloaded like this.

- Tripping over cables
  Cause – trailing wires present a tripping hazard.
  Prevention – the cables need to be managed by sinking into floors or by covering by floor covering.
- Electrocution
  Cause – faulty equipment, users tampering with the inside of computers or users spilling drinks over computer equipment.
  Prevention – any malfunctioning equipment must not be used and should be reported to the technician. Drinks should be kept away from computers.
- Heavy equipment falling
  Cause – equipment not positioned properly.
  Prevention – ensure that the equipment has enough space to be safely positioned.

QUESTIONS C

1 The use of ICT systems has been associated with a number of health problems.
   a State **three** health problems that have been associated with the prolonged use of ICT systems. (3 marks)
   b In order to prevent health problems caused by the use of computers, some actions can be taken. Describe **six** such actions that can be taken to avoid the health problems you have identified in part (a) happening. (6 marks)

2 An employee who spends much of their time at a keyboard typing in orders at high speed is worried about RSI.
   a What do the initials RSI stand for? (1 mark)
   b Give one of the symptoms of RSI. (1 mark)
   c Write down **two** precautions that the employee can take to minimize the chance of contracting RSI. (2 marks)

3 Copy the table and tick (✓) the correct column to show whether each of the following statements about health risks in using ICT is true or false. (5 marks)

	TRUE	FALSE
The continual use of keyboards over a long period can give rise to aches and pains in the hands, arms and wrists	☐	☐
RSI stands for repeated stress injury	☐	☐
Wrist rests and ergonomic keyboards can help prevent RSI	☐	☐
Back ache can be caused by slouching in your chair when using a computer	☐	☐
Glare on the screen can cause RSI	☐	☐

4 Some people misuse email facilities. For example, spam is an annoyance for many computer users.
   a Explain what is meant by spam. (1 mark)
   b Describe how it is possible to reduce the amount of time dealing with spam. (2 marks)

5 Many people get sent phishing emails.
   a What is meant by phishing emails? (2 marks)
   b Explain the dangers in these phishing emails. (2 marks)

! Revision Tip

Always be guided by the mark scheme to decide how much to write. For example, if you are asked to describe an advantage or disadvantage for one mark, then just a brief statement would be enough. If two marks were allocated, then you would be required to supplement this with further detail or an appropriate example.

Always think out your answer before you start writing it. You need to ensure you make your answer clear, so you need a little time to think about it.

Advantages (sometimes called benefits) and disadvantages are very popular questions. When covering a topic for revision, it is a good idea to list advantages and disadvantages where appropriate.

You need to be clear about the health problems (i.e., what they are called), the symptoms (i.e., how they affect your body) and what can be done to help to prevent them.

Activity 6.3

Repetitive strain injury (RSI) has become a major worry for those people who use computers continually throughout their working day.

You are required to use the Internet to find out more about this condition. You need to find out:
» What are the symptoms?
» Can you make it better?
» What is the likelihood of getting it?
» What can you do to prevent it?

Test yourself

The following notes summarize this topic. The notes are incomplete because they have words missing. Using the words in the list below, copy out and complete the sentences A to P, underlining the words that you have inserted. Each word may be used more than once.

RSI	files	headaches	eye-tests	scanner
download	blogs	spam	wikis	hacking
blinds	phishing	firewall	pharming
teleworking	back ache	anti-virus

A To protect data from deliberate damage caused by hackers illegally gaining access to a computer network via the Internet, a _____ should be used.

B In order to prevent viruses entering an ICT system, _____ software should be used to search for and destroy viruses.

C Users should be told not to open _____ attached to emails unless they know who they are from.

D Users should also be told not to _____ music and games illegally off the Internet from file sharing sites.

E _____ is caused by typing or using a mouse over a long period of time.

F The symptoms of eye strain include blurred vision and _____.

G Working in cramped conditions and not adopting the correct posture when using computer can lead to _____

H Adjustable _____ should be used on windows to prevent glare on the screen, and the screen should also be kept free from glare from lights.

I It is important to have regular _____ and use glasses or contact lenses when working with computers, if needed.

J Unauthorized use of an ICT system with a view to seeing or altering the data is called _____.

K _____ means sending emails which falsely say they are from banks or other organizations in order to trick people into revealing their banking or credit card details.

L _____ means working from home by making use of computers and communications equipment.

M It is possible to use special software to filter out emails that are unasked for, which are popularly called _____.

N _____ is where program code is deposited onto a user's computer (by a virus/Trojan) or onto a server in the case of a network.

O Web sites that are created by an individual with information about events in their life, videos, photographs, etc., are called _____.

P Web pages that can be viewed and modified by anyone who has a web browser are called _____.

Important note

Here are some of the words you will see being used in examination questions and what they mean:

State/List – single words or phrases are usually enough.
Describe – needs a larger answer to show understanding.

It is important to be able to distinguish between health and safety issues.

Health (person) – RSI, headaches, eye strain and backache.
Safety (incidents) – fire, electrocution, tripping, etc.

REVISION QUESTIONS

1 a What do the initials RSI stand for? [1]

b RSI is a health problem that may be caused by prolonged computer use.
Describe how RSI is caused. [2]

c List **one** precaution that a computer user can take to minimize the chance of contracting RSI. [1]

2 Here is a list of health problems. Write down the names of those that can be caused by prolonged computer use: [4]

Back ache
Toothache
Sprained ankle
RSI
Eye strain
Headaches

3 People who work with computers for long periods may experience some health problems. These health problems include eye strain and RSI.

a Give the names of **two** health problems other than eye strain and RSI that a user may experience. [2]

b Explain **two** things a user should do in order to prevent future health problems when sitting in a chair at a desk and using a computer. [2]

4 Describe **one** way in which ICT is used in manufacturing in order to reduce the number of people needed to produce goods. [3]

5 There have been a lot of changes in the pattern of employment due to the increased use of ICT.
Tick **three** boxes that give sensible reasons why the pattern of employment has changed with increased use of ICT. [3]

Tick **three** boxes only

Homeworking/teleworking is more popular ☐
Employees are more likely to work more flexibly ☐
It has brought about a huge rise in employment especially among factory workers ☐
Training and retraining are needed regularly ☐
Workers are generally less skilled than they were ☐

EXAM-STYLE QUESTIONS

1 There are many examples of abuses of the Internet.
  Explain what is meant by the following terms.   *[2]*
  a Pharming
  b Phishing

2 Spam causes annoyance to computer users.
  Explain what is meant by spam and give **one** example of why it causes annoyance to
  computer users.   *[2]*

3 There are a number of health and safety issues caused by the use of computers.   *[4]*
  a   Describe **two** health issues that are caused by computer use.
  b   Describe **two** safety issues that are caused by computer use.

4 Explain what is meant by the following terms   *[3]*
  a Blog
  b Social networking site
  c Wiki

7 The ways in which ICT is used

ICT is used in almost every area of our lives, so in this chapter you will be looking at the range of ICT applications in everyday life. You will also be looking at the impact of developments in ICT which shape the way we go about our lives.

The key concepts covered in this chapter are:
▶▶ Communication applications
▶▶ Interactive communication applications (e.g., blogs, wikis, and social networking sites)
▶▶ Data handling applications (e.g., surveys, address lists, tuck shop records, clubs and society records, school reports, and school libraries)
▶▶ The differences between batch processing, online processing, and real-time processing
▶▶ Online booking systems
▶▶ Applications of ICT in finance departments
▶▶ Measurement applications (e.g., scientific experiments, electronic timing, and environmental monitoring)
▶▶ Control applications
▶▶ Modelling applications
▶▶ Applications of ICT in school management
▶▶ Applications of ICT in banking
▶▶ Applications of ICT in medicine
▶▶ Applications of ICT in libraries
▶▶ The use of expert systems
▶▶ Applications in the retail industry

Communications applications

Most people use ICT to communicate in some way with others and this can be in a personal capacity or a work-related capacity. In later chapters you will be preparing documents, web sites, presentations, etc., that allow you to communicate effectively with others.

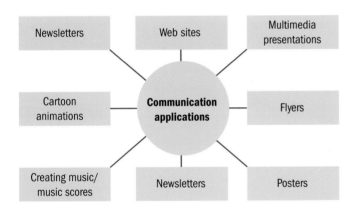

Cartoon animations
Animations can bring the most boring subject to life and they are used extensively in web sites and products to help children learn. They are also used in web sites to sell goods and services.

You have probably heard of Flash, which is software you can use to create animations. Simple animations often make use of key frames. Key frames are the starting frame and the finishing frame in an animation. For example, if you had a simple animation of one shape such as a square changing into a different shape such as a circle then key frames would be:

▶▶ the start frame of the square
▶▶ the end frame of the circle.

Here are the steps you would need to take to move a red ball across the screen from left to right:

1 Draw the red ball
2 Erase it
3 Draw the red ball a little to the right
4 Erase it
5 Draw the red ball a little further to the right
6 Repeat steps 3 to 5 enough times
7 Draw the red ball in its final position.

Using Flash (and other brands of animation software), when you want an object to move across the screen, you need only to define the key frames (i.e., drawn in steps 1 and 7) where something changes. You leave it to Flash to put in the frames in between. This is called tweening and it enables simple animations to be produced quickly.

Creating music/music scores
Musicians, composers and music producers make extensive use of ICT in the following ways:

▶▶ Sequencers can be used. Sequencers are hardware or software used to create and manage electronic music and they include such things as drum machines and music workstations that allow musicians to create electronic music.

- Sound wave editors are used which are software that allows the editing of sound waves. Using the software sound waves can be edited, cut, copied, and pasted, and also have effects like echo, amplification, and noise reduction applied.
- MIDI (Musical Instrument Digital Interface) can be used. MIDI enables a computer and a musical instrument to communicate with each other. For example, when a keyboard is played, the music is transferred using MIDI to the computer where it can be edited if needed and stored. The process can be reversed by the music stored on the computer being fed back and using MIDI it can play the music back on the keyboard or other instrument.
- Notators (music composition software) can be used. A notator is a piece of software that allows you to compose your own music and you do this by entering notes into the computer via:
  - the keyboard
  - a MIDI system
  - or scanning a piece of music on paper using a scanner.

Using ICT equipment to edit sound in a studio.

Interactive communication applications

ICT systems are used by organizations and individuals to communicate with each other. There are many examples of ways people and organizations can communicate effectively and these include:

- blogs
- wikis
- social networking web sites.

All of these were covered in the Chapter 6.

Data handling applications

Many ICT systems perform simple data handling tasks. In these systems the data is input into the computer into a simple database structure and it can then be processed in some way and the results output in the form of reports.

Surveys

Surveys are used to collect information by asking lots of different people the same set of questions. Because of the huge amounts of data collected, the questionnaires might use OMR (optical mark recognition) for the input of data. The completed questionnaires can be read at high speed by the optical mark reader, which saves time and money. Once the data has been entered it can be stored and processed to produce statistics or graphs and charts.

Tuck shop records

ICT systems are used to keep records of stock and takings (i.e., the money coming in when the goods are sold) and such systems might also make use of bar code readers. Spreadsheet software can be used to model the income and expenditure and perform "what if?" calculations to see how profits from the tuck shop can be improved.

Clubs and society records

Clubs and societies have income and expenditure which needs to be recorded, and computers with appropriate software can be used for this. They need to also keep membership details such as the contact details of their members. They will have to keep details of membership renewals and use mail merge to send membership renewal reminders out to members.

Members fill in a form and the details are input into a database for storage on the computer.

The differences between batch processing, online processing, and real-time processing

Data can be processed by the computer in a number of different ways and these are called:

- batch processing
- online processing
- real-time processing.

Batch processing

Batch processing is used when a particular job needs to be done in one go rather than in a number of parts. All the relevant data is collected/batched and processed together. Batch processing has the advantage that once all the inputs are ready and the program has been selected, the computer can just get on with the job without any human intervention.

Batch processing is ideal for:

▸▸ producing attendance statistics from attendances recorded on OMR forms
▸▸ producing bills for water, gas, telephone, and electricity companies
▸▸ producing monthly bank or credit card statements
▸▸ marking multiple-choice examination papers.

Online processing

With online processing, the system is automatically updated when a change (called a transaction) is made. This means that the system always contains up-to-date information. An airline booking system would use online processing because, as seats are sold, the number of seats available would need to be reduced to prevent double booking. Seats are reserved whilst a customer is booking them and the files are then updated when the seat purchase is finalized and paid.

Online processing is ideal for:

▸▸ booking systems (for airline, holiday, concert, and theatre tickets)
▸▸ ordering goods over the Internet
▸▸ banking transactions such as withdrawing money from an ATM, making a payment using Internet banking, etc.

Real-time processing

Real-time processing is an example of online processing and is used where it is essential that the results of processing are obtained immediately without any delay. The system responds immediately and alters the system in some way. Sensors are used to collect the data and the processor sends signals to actuators which will turn devices on or off. The computer then processes the data and changes the output in some way, which in turn will change the next lot of data received by the sensors. Real-time systems are used for control.

Real-time processing is ideal for:

▸▸ flood warning systems
▸▸ autopilots for aircraft
▸▸ computer games
▸▸ traffic light control
▸▸ process control in factories (i.e., making steel, chemical plants, etc.)
▸▸ controlling robots
▸▸ control of the environment in a greenhouse.

Revision Tip

In data processing systems (i.e., the type used by businesses) the processing is always batch or online. Real-time systems are used mainly for control where sensors are used and there needs to be an immediate response.

Online booking systems

The main feature of all online booking systems is that they all use online processing. This means that while the transaction is taking place and the user is entering their details, the seats are saved for them. If they choose not to go ahead with their purchase, the seats held are released for sale to others. This immediate updating of files prevents double bookings.

Online booking systems can be used to book:

▸▸ holidays
▸▸ flights
▸▸ train tickets
▸▸ cinema seats
▸▸ theatre seats
▸▸ tickets for sporting events.

The steps for online booking are as follows:

▸▸ Find the booking web site using a search engine or type in the web address.
▸▸ Search the online booking database using dates, times, etc.
▸▸ Make your selection (in many cases you can choose where you sit from a plan).
▸▸ The seats are now held for you so no-one else can book them.
▸▸ Enter your details (names, addresses, etc.).
▸▸ Select a payment method (e.g., credit card, debit card, etc.) and enter the card details.
▸▸ A confirmation appears on the screen telling you that the seats/holiday, etc., have been successfully booked.
▸▸ A confirmation email is sent to you, which in many cases can be printed out as it acts as your ticket.

Advantages of online booking:

▸▸ You can book from the comfort of your home at any time.
▸▸ There is more time to look for holidays, flights, etc., than when at a travel agents.
▸▸ You can make savings for flights/holidays when you book direct as there is no travel agent commission to pay.
▸▸ You can read reports from people who have been on the same holiday, seen the same concert, etc.
▸▸ There is no need to pick up tickets as you often print these yourself.

Disadvantages of online booking:

▸▸ You have to enter credit/debit card details and these may not be kept safe.
▸▸ People could hack into the site and know you were away and burgle your house.
▸▸ There is no personal service like at a booking agent.
▸▸ You could easily enter the wrong information and book the wrong flights or performance on the wrong day.

QUESTIONS A

1 Put a tick in the column which best describes the type of processing used in the following applications. (*6 marks*)

	Online	Batch	Real-time
Ordering goods using the Internet			
Processing the results from a questionnaire			
Preparing water bills automatically			
Controlling the flight of a plane automatically			
Processing a payroll and printing wage slips			
Control of a patient life support system in a hospital			

2 Describe **two** ways in which the use of ICT has made it easier for musicians to compose and create music. (*2 marks*)

3 Explain the differences between batch and real-time processing. (*2 marks*)

4 A web site is being created that enables customers to place orders for groceries and get them delivered to their home.
   a Give the name of the type of processing which is appropriate. (*1 mark*)
   b Explain why the type of processing you have named in your answer to part (a) is appropriate. (*1 mark*)

Applications of ICT in finance departments

Many organizations have finance departments to keep track of money coming into and going out of the business. The following ICT systems are used in finance departments:

▸▸ billing systems
▸▸ stock control systems
▸▸ payroll systems.

Billing systems

As the name suggests, billing systems are used to send bills to customers who will either pay online or send a cheque back by post with the payment slip. Because of the huge volume of bills that are produced by water, gas, and electricity companies, and other companies such as credit card companies, the processing is highly automated.

Many billing systems use batch processing to produce the bills for each customer and also to process the payments made by post using cheques. Many billing systems offer the customer ways of paying their bill online using a credit or debit card. Many companies send an electronic bill rather than a paper bill, which reduces costs and is less wasteful in terms of resources.

Stock control systems

Most organizations carry stock. They need financial systems that will issue orders when stocks of items run low and pay for them when they are delivered.

In shops the stock control system is linked to the Point of Sale (PoS) terminals (computerized tills). When an item is sold at the PoS terminal the number of that particular item in stock is reduced by one. This means that the computer knows how many items are in stock. Once the number of items has fallen below a certain level, the computer system will automatically order more stock from the supplier. This means that stores should not run out of fast-selling items.

Payroll systems

All organizations employ staff who need to be paid and the system for paying them is called a payroll system.

Some staff are paid monthly and others are paid weekly according to the number of hours they work. Some staff may get paid overtime and their hourly rate may increase if they work nights, weekends, etc. Some employees may have part or all of their pay dependent on commission or bonuses. As you can see, payroll can get quite complicated.

There are also deduction such as tax, insurance, and pension. Because of the security and cost problems of dealing with the transfer of large amounts of cash, many organizations use electronic funds transfer (EFT) to transfer the money between the organization's bank account and the employees' accounts. They can also use the system to make payments of income tax and insurance contributions to the authorities that have been deducted from employees' pay.

Data capture, checking and type of processing

There are lots of ways in which data is captured and these include:

▸▸ Employees clock in and clock out using a card or key – this records the hours they have worked and these are input into the computer directly.
▸▸ Employees fill in timesheets that are read automatically – sometimes these use techniques such as OMR (optical mark recognition) where the employee shades in boxes on a form. Other methods use OCR (optical character recognition), where

the reader is able to read numbers and letters the employee fills in on a form.

▸▸ Employees fill in timesheets that are input manually using a keyboard.

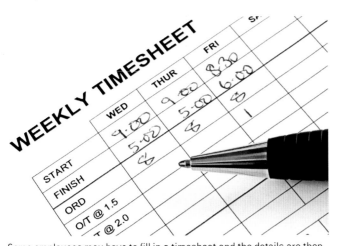

Some employees may have to fill in a timesheet and the details are then typed in using a keyboard.

Processing timesheets

Both types of form can be collected and batched together to be read automatically at the end of the week or month. They are then input into the computer, which processes them in one go. This type of processing is an example of batch processing.

Output from a payroll system

After the pay has been calculated a document called a payslip is printed. As most payrolls are processed in a batch, the payslips are also printed in a batch. The payslip provides confirmation of the hours worked the pay, the tax, insurance, and other deductions and the net pay.

Measurement applications

Computers can be used in conjunction with sensors to measure physical quantities such as temperature, pressure, light intensity, infra-red radiation, etc. Because most of these quantities are analogue, an analogue to digital converter is needed to change the data into digital data that can be processed and stored by a computer.

Monitoring

Monitoring involves taking readings/measurements regularly or continuously over a period of time using sensors. There are three things that can then happen:

▸▸ They can be stored for later – the readings can be stored and then output in some way, such as graphically.

▸▸ Set off alerts – they could be used to issue a warning sound such as the beeper if the heart rate of a patient being monitored changes.

▸▸ Auto control – the readings can be used to control a device in some way. For example, the readings sent from a temperature sensor could be used to turn an air-conditioning unit on or off.

Monitoring can be used in science lessons for recording temperature, light, force, etc. In geography lessons monitoring can be used to record the weather. It can also be used in society for monitoring traffic flow, monitoring pollution, keeping track of climate change, for electronic timing, etc.

The main features of monitoring are:

▸▸ The readings are taken automatically – there is no need for a human to be present. This means that it is much cheaper than employing a person to do this.

▸▸ You can set the period over which the monitoring takes place – this is the total time over which the readings will be collected. You can also set the system to take readings continuously such as in a flood warning system.

▸▸ You can set how often the readings are taken. For example, in an experiment to investigate the cooling of boiling water, you might decide to set the frequency with which readings are taken to be every minute.

▸▸ The sensors can be put in remote locations – you can put them anywhere in the world and the data can be sent back wirelessly and even using satellites.

▸▸ The sent data can be stored and processed by a computer.

▸▸ The data can be analysed (you can do calculations such as work out the mean, mode median, range, etc.), and draw graphs and charts. The data can be imported into software such as a spreadsheet package.

Sensors

Sensors are used to detect and measure physical quantities. Here are some examples of sensors:

▸▸ Temperature/heat sensors – can be used in school experiments such as investigating the cooling of a hot drink in different thicknesses of cardboard cup. Heat sensors can be used to control a heating/cooling system in a home or classroom/greenhouse.

▸▸ Light sensors – detect the brightness of light. Can be used to see how light levels affect the growth of a plant. They can be used to control lights that come on automatically when it goes dark in a greenhouse.

▸▸ Sound sensors – measure the loudness of a sound. They can be used in noise disputes (e.g. monitoring sound at airports).

▸▸ Pressure sensors – barometric pressure sensors measure air pressure; other pressure sensors measure depth of liquid or something pressing on them.

▸▸ Humidity sensors – these measure the moisture in the air in greenhouses or art galleries.

▸▸ Passive infra-red sensors (PIRs) – these are the sensors used in schools and homes to detect movement. They can be used in burglar alarms and also to turn lights on/off automatically in rooms when a person walks in/out.

Advantages and disadvantages of monitoring using ICT

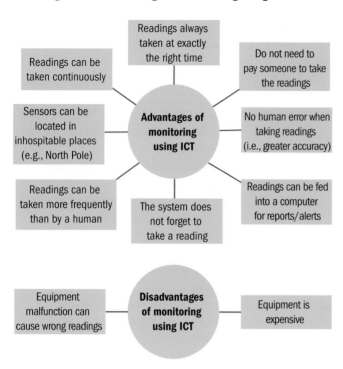

Readings can be taken continuously

Readings always taken at exactly the right time

Do not need to pay someone to take the readings

Sensors can be located in inhospitable places (e.g., North Pole)

Advantages of monitoring using ICT

No human error when taking readings (i.e., greater accuracy)

Readings can be taken more frequently than by a human

The system does not forget to take a reading

Readings can be fed into a computer for reports/alerts

Equipment malfunction can cause wrong readings

Disadvantages of monitoring using ICT

Equipment is expensive

Environmental monitoring

Environmental monitoring is used to collect data for weather forecasts, to test water quality in river and streams, to collect data about air pollution, etc. Sensors take the readings automatically and these readings can be saved on removable media or the data can be sent back to the computer automatically using a wireless signal.

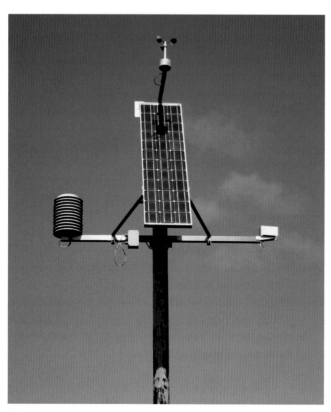

This weather station monitors wind speed, temperature, atmospheric pressure, and humidity.

Control applications

Data from sensors can be used to control devices. For example, the data from temperature sensors are used to turn heaters on or off to maintain a constant temperature. Most household appliances you find in the home such as washing machines, air conditioning units, kettles, etc., use some form of control, as you will see later in this chapter.

Using a sequence of instructions to control devices

In control systems it is necessary to give a series of commands for the system to obey.

There is an educational package for teaching about control commands called LOGO and it is used to move a turtle or cursor around the screen. The turtle or cursor can be instructed to leave lines to trace its path on the screen. For example, the following set of commands can be used to move an arrow on the screen of a computer. When the arrow moves, it leaves a line.

> FORWARD distance
> LEFT angle
> RIGHT angle

Hence, using the commands FORWARD 5 would move the arrow forward 5 units and LEFT 90 would turn the arrow left through an angle of 90°.

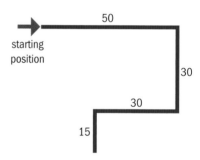

Here is a shape that was drawn on the screen with the numbers representing the lengths of lines.

The list of instructions that would draw this shape is as follows:

> FORWARD 50
> RIGHT 90
> FORWARD 30
> RIGHT 90
> FORWARD 30
> LEFT 90
> FORWARD 15

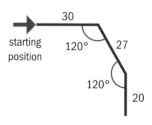

Here is a more complex shape that makes use of angles other than right angles.

The set of instructions to draw this shape on the screen are:

FD 30
RT 60
FD 27
RT 60
FD 20

Repeating sets of instructions: instructions can be repeated using the following commands:

REPEAT n
ENDREPEAT

The instructions to be repeated a set number (i.e., n) times are placed in between these two commands.

For example a square could be drawn using REPEAT n using the following commands:

REPEAT 4
FORWARD 20
RIGHT 90
ENDREPEAT

There are also some other commands:

PENDOWN (this instruction puts the pen down so that a line is ready to be drawn)
PENUP (this instruction raises the pen so the pen can be moved along without drawing a line)

Simple control applications

Simple control systems use data from sensors.

Security light system – uses a PIR (passive infra-red) sensor to sense movement. As soon as the sensor detects movement the system turns the light on. After a period of time the system turns the light off.

PIR sensors are used to detect intruders and sound an alarm siren in a burglar alarm system.

A burglar alarm – works in a similar way to the security light using PIRs as the input into the system. This time the output device is a bell or siren that sounds when the alarm is on and movement is detected.

Automatic cookers – temperature sensors relay temperature data back to the microprocessor to control the operation of the heaters. The microprocessor also controls when the oven is to come on and how long it stays on for.

Microwave ovens – used to control the strength of the microwaves and their duration. Some ovens and microwave ovens can read the bar code on food packaging, which tells the oven how to cook the food. Once read, the bar code gives the oven control information such as the correct temperature/power and the time needed to cook the food.

Computer-controlled greenhouses

In order to grow plants successfully they need perfect growing conditions. Computer control can be used to monitor the conditions and keep these conditions constant.

The sensors used to collect the input data include:

▶▶ light
▶▶ moisture
▶▶ temperature.

The output devices that are controlled by the computer include:

▶▶ lamps (to make the plants grow faster)
▶▶ heater
▶▶ motor to turn the sprinkler on/off to water the plants
▶▶ motor to open or close the windows (to cool the greenhouse down if it gets too hot).

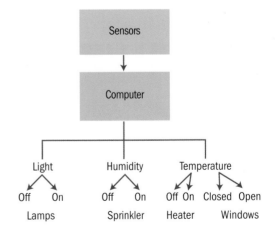

The microprocessor compares the sensor reading to a preset value.

The microprocessor sends signals to actuators.

If temperature reading > preset value, the microprocessor switches the heater on and opens the windows.

If temperature reading < preset value the microprocessor switches the heater on and closes the windows.

If light reading < preset value the microprocessor switches the lights on.

If light reading > preset value the microprocessor switches the lights off.

If humidity reading < preset value the microprocessor switches the sprinkler on.

If humidity reading > preset value the microprocessor switches the sprinkler off.

Tomatoes need ideal growing conditions to maximize the crop.

Sprinkler system in a commercial greenhouse.

Robotics

Robots have been widely used in manufacturing for years, especially for painting and welding in car factories. Robots are also used for picking and packing goods in large warehouses. Robots have been developed that will do some of the tasks humans hate to do such as mowing the lawn or vacuuming the floors.

Robots have been developed for use on farms and these robots can perform a variety of farm tasks such as planting, weeding inbetween crops, crop spraying and picking crops.

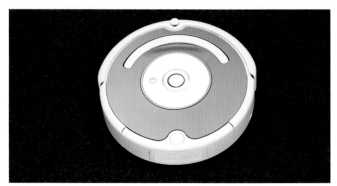

Robots will eventually be seen in all homes. This vacuuming robot is already in the shops.

There are robots available for the home that will wash floors, clean gutters, and clean swimming pools. The robots that are available at the moment in the home are usually capable of performing one task. In the future you will probably buy a single multifunctional robot capable of carrying out a range of different tasks.

Mowing the lawn is a chore for many people, so this robot lawnmower is a useful device.

Advantages and disadvantages of control systems/robots

Compared to manual control, computer-based control systems or robots have the following advantages and disadvantages.

Advantages:

▸▸ Can operate continuously, 24 hours per day and 7 days per week.
▸▸ Less expensive to run as you don't have to pay wages.
▸▸ Can work in dangerous places (e.g., a robot to remove a bomb).
▸▸ Can easily change the way the device works by re-programming it.
▸▸ More accurate than humans.
▸▸ Can react quickly to changes in conditions.

Disadvantages:

▸▸ Initial cost of equipment is high.
▸▸ Equipment can go wrong.
▸▸ Fewer people needed so leads to unemployment.

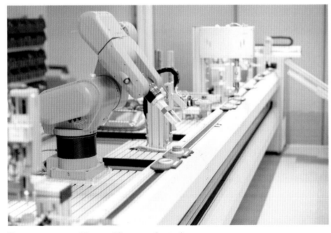

This robot is used by the army to make bombs safe.

Computer control is used in manufacturing.

QUESTIONS B

1 A turtle that draws on paper uses the following instructions.

FORWARD *n*	Move *n* cm forward
BACKWARD *n*	Move *n* cm backwards
LEFT *t*	Turn left *t* degrees
RIGHT *t*	Turn right *t* degrees
PENUP	Lift the pen off the paper
PENDOWN	Lower the pen onto the paper

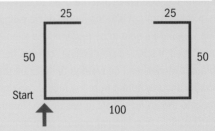

Write a set of instructions that makes the turtle draw the shape shown above.
Assume that the pen is down at the start. *(4 marks)*

2 Computer control is used to control growing conditions in a greenhouse. For example, a temperature sensor will turn on a heater if the temperature inside the greenhouse gets too cold.
  a Give the names of **two** other sensors that could be used in the greenhouse and for each one describe why it is needed. *(4 marks)*
  b Give **two** advantages in using ICT to monitor and control the growing conditions in the greenhouse. *(2 marks)*
  c Describe **one** disadvantage in using monitoring and control to control the growing conditions in the greenhouse. *(1 mark)*

3 Give the names of **three** different output devices that can be controlled by a computer. *(3 marks)*

4 A home weather station consists of a base station that contains the processor and the display. Sensors are also included that are placed outside the house. Readings from the sensors are relayed back to the base station, which processes the data and produces weather information that is displayed on the screen.
  a Give the names of **two** different types of sensor that could be used with this system. *(2 marks)*
  b Describe **one** method by which the data can get from the remote sensors to the base unit that is situated inside the house. *(2 marks)*
  c Once the data has been sent to the base unit it is processed and the information is output. Describe **one** way that the weather information is output from the system. *(2 marks)*

Modelling applications

Spreadsheet software can be used to construct models and can be used to produce financial models such as the money coming into and going out of a school tuck shop or models on personal finance to help you manage your money. Modelling means producing a series of mathematical equations that are used to mimic a real situation. When values are put into the model or we exercise the model in some way, we are said to be performing a simulation. There are many types of specialist modelling software from games to flight simulators. Models can be created to describe the flow of traffic at junctions and the output from the model can then be used to issue controls to the traffic signals to ensure that the traffic flows as smoothly as possible thus keeping queues to a minimum.

Modelling can also be applied to graphics. For example, you can produce a computer model of a building in 3D using a plan of the building in 2D. Architects can make changes on the plan and then see the effect of the changes on the 3D view.

The advantages and disadvantages of using simulation models
Some advantages:

- Cost – it can be cheaper to use a model/simulation. For example, car engineers can use a computer to model the effect on the occupants during a crash and this is cheaper than using real cars with crash test dummies.
- Safer – flight simulators can model the effect of flying a plane in extreme situations. Extreme situations might include landing without the undercarriage coming down, or landing with only one engine working, and so on. It would be far too dangerous to expect a pilot to try these in real life.
- It can save time – global warming models can be set up to predict what the likely effects of global warming will be in the future.
- It is possible to experience lots more situations – pilots using simulations can experience all sorts of extreme weather conditions such as sand storms, hurricanes, smoke from volcanic eruptions, and so on. These would be almost impossible to experience any other way.

Some disadvantages:

- The differences between simulation and reality – there will always be some difference between a model/simulation and reality. No model or simulation can ever be perfect because real life can be so complicated.
- The accuracy of the rules and variables – the person designing the model may have made mistakes with the rules or the variable data.

- Some situations are hard to model – some situations are difficult to model because some aspects of the model are often open to interpretation. For example, experts on the subject may disagree about the rules that apply.

Applications of ICT in school management

There are many ways in which ICT is used to help with the administration and management of schools.

School registration
School registration systems are something you will be familiar with. You will know that it is important for the school to keep a record of who is on the premises and who is not.

Computer-based methods of registration
Any ICT system used for student registration in schools or colleges should:

- capture student attendance accurately
- capture the student attendance automatically
- be very fast at recording attendance details
- as far as possible avoid the misuse of the system
- enable not only morning and afternoon attendance to be recorded but also to record attendance at each lesson and enable attendance patterns for individuals and defined groups to be worked out
- be relatively inexpensive
- work with other ICT systems used in the school, such as the system for recording student details.

Optical mark recognition (OMR)
Optical mark recognition works by the teacher marking a student's attendance by shading in boxes using a pencil. The forms are passed to the administration office where they are collected and batched together and processed automatically using an optical mark reader.

Advantages of optical mark recognition:

- Frees up staff from entering attendance marks manually using a keyboard.
- The OMR reader is cheap.
- Reader can be used for other purposes such as reading multiple-choice answer sheets.

Disadvantages of optical mark recognition:

- Registration is not done in real time – if a student came in halfway through the morning, this system would not record this.
- Registers need to be passed manually to the administration staff.
- Registers are easily altered by students.
- If the forms are folded or damaged, they are rejected by the reader.

ARC

ABSENCE REPORT CARD **CARD NO.**

[0]	[0]	[0]	[0]	[0]	[0]
[1]	[1]	[1]	[1]	[1]	[1]
[2]	[2]	[2]	[2]	[2]	[2]
[3]	[3]	[3]	[3]	[3]	[3]
[4]	[4]	[4]	[4]	[4]	[4]
[5]	[5]	[5]	[5]	[5]	[5]
[6]	[6]	[6]	[6]	[6]	[6]
[7]	[7]	[7]	[7]	[7]	[7]
[8]	[8]	[8]	[8]	[8]	[8]
[9]	[9]	[9]	[9]	[9]	[9]

INSTRUCTIONS
Use only pencil when
completing this form.
Mark like this ⊬

Student name	Date	Present	Code
		["]	[L] [M] [N] [P] [T] [H] [G] [J] [F]
		["]	[L] [M] [N] [P] [T] [H] [G] [J] [F]
		["]	[L] [M] [N] [P] [T] [H] [G] [J] [F]
		["]	[L] [M] [N] [P] [T] [H] [G] [J] [F]
		["]	[L] [M] [N] [P] [T] [H] [G] [J] [F]
		["]	[L] [M] [N] [P] [T] [H] [G] [J] [F]
		["]	[L] [M] [N] [P] [T] [H] [G] [J] [F]
		["]	[L] [M] [N] [P] [T] [H] [G] [J] [F]
		["]	[L] [M] [N] [P] [T] [H] [G] [J] [F]
		["]	[L] [M] [N] [P] [T] [H] [G] [J] [F]
		["]	[L] [M] [N] [P] [T] [H] [G] [J] [F]
		["]	[L] [M] [N] [P] [T] [H] [G] [J] [F]
		["]	[L] [M] [N] [P] [T] [H] [G] [J] [F]
		["]	[L] [M] [N] [P] [T] [H] [G] [J] [F]
		["]	[L] [M] [N] [P] [T] [H] [G] [J] [F]
		["]	[L] [M] [N] [P] [T] [H] [G] [J] [F]
		["]	[L] [M] [N] [P] [T] [H] [G] [J] [F]
		["]	[L] [M] [N] [P] [T] [H] [G] [J] [F]
		["]	[L] [M] [N] [P] [T] [H] [G] [J] [F]
		["]	[L] [M] [N] [P] [T] [H] [G] [J] [F]

An OMR form which is used to enter student registration information into the Schools Information Management System.

Smart cards

Smart cards look like credit cards and they contain a chip that can be used to hold certain information. Smart cards hold more information than cards containing only a magnetic stripe and can be used in schools in the following ways:

▸▸ for registration of students
▸▸ for monitoring attendance at each lesson
▸▸ for payment for meals in the school canteen
▸▸ for access to the school site, buildings, and rooms to improve security
▸▸ for access to certain facilities such as the computer network, photocopier, etc.
▸▸ to record borrowing and return of school library books, digital cameras, musical instruments, etc.

Swipe cards

Students are given a swipe card that they use for registration purposes by swiping the card using a card reader. Swipe cards are plastic cards with a magnetic stripe containing a limited amount of data on it. The swipe card is used to identify the student to the registration system and some other systems such as the library system and the school meals system. The same card can be used for access to school buildings.

Advantages of swipe cards:

▸▸ The cost of the cards and the readers is low compared to other methods.
▸▸ Readers can be made that are almost vandal proof.

Disadvantages of swipe cards:

▸▸ Cards are often lost or forgotten, meaning that students have to be registered using a keyboard.
▸▸ Students cards can be swiped in by someone else.

Biometric methods

Biometric methods provide a fast and easy way of recording student attendance. Biometric methods make use of a feature of the human body that is unique to a particular person in order to identify them. Biometric methods include:

▸▸ fingerprint recognition
▸▸ retinal scanning.

Advantages of biometric methods:

▸▸ There is nothing for a student to forget like a card.
▸▸ You have to be there to register so no-one else can do it for you and it cannot be altered by students.
▸▸ Performed in real time so the system knows exactly who has registered and when.

Disadvantages of biometric methods:

▸▸ Biometric systems are expensive.
▸▸ There are privacy issues. Some people object to fingerprinting systems.
▸▸ People can vandalize the readers.

School management systems

There are many different pieces of information stored by a school – the student records and the records of attendance are just two. By integrating all the systems, it is possible to extract the data needed from the system in the form of reports. A report might be a list of all those students with 100% attendance so they can be given a certificate. Exam, homework, and test results can be entered into the system and the system can be used by teachers to enter reports for each student. It is also possible to allow some access to the system by parents to enable them to review the progress of their child.

School management systems are ICT systems that supply school managers and staff with information that can help them make decisions. For example, the attendance system might produce information about those students for whom attendance is bad. The system might produce a report showing when they are absent to see if there is a pattern. The senior teachers can then take action.

Here are some other ways school management systems can be used:

▸▸ To work out how many students will be in the new intake and to allocate them into tutor groups.

- ▶▶ To decide whether a new teacher should be employed.
- ▶▶ To work out the best way of allocating teachers and classrooms.
- ▶▶ To decide on how best to spend the training budget to keep teachers up-to-date.

The main advantages in using school management systems are:

- ▶▶ They reduce the workload for teachers in the classroom and in the school office.
- ▶▶ They can provide up-to-date information for parents.
- ▶▶ They can support decision making for school managers.
- ▶▶ They can tackle truancy effectively.
- ▶▶ They can be used to plan timetables.

The main disadvantages in using school management systems are:

- ▶▶ The software is expensive to buy.
- ▶▶ Student data is personal, so there must be no unauthorized access.
- ▶▶ Software is complex, so all staff need training.

Many schools use ICT-based school management systems for student registration, keeping student records and producing reports.

A pupil record from the Schools Information Management System. Notice that included in the details is a photograph of the pupil.

Applications of ICT in banking

Many of the facilities offered by banks such as the use of credit/debit cards and online banking would not be possible without the use of ICT. Banks use ICT in many ways which are outlined below.

Cheque clearing

Cheques are still used and millions of them go through a process called cheque clearing each day. Check clearing uses input methods that use magnetic ink characters printed on the cheque.

With cheque clearing, the numbers are printed onto the cheque in a special magnetic ink, which can be read at very high speed by the magnetic ink character reader. Most of the data (cheque number, bank sort code, and account number) are pre-printed onto the cheque but the amount is not known until the person writes the cheque.

Bank cheques are read at high speed during the clearing process.

When the cheque is presented for payment, the amount the cheque is for is then printed onto the cheque, again in magnetic ink. All the cheques are batched together at a centre operated by all the banks called a clearing centre. All the cheques are read and processed in one go by the machine, so this is an example of batch processing.

Advantages of magnetic ink character recognition include:

- ▶▶ Accuracy – the documents (usually cheques) are read with 100% accuracy.
- ▶▶ Difficult to forge – because of the sophisticated magnetic ink technology used, it would be difficult to forge cheques.
- ▶▶ Can be read easily – cheques are often folded, crumpled up, etc. Methods such as OCR or OMR would not work with these. Magnetic ink character recognition uses a magnetic pattern, so this is unaffected by crumpling.
- ▶▶ Speed of reading – documents can be read at very high speed and this is particularly important for the clearing of cheques.

Disadvantages of magnetic ink character recognition include:

- ▶▶ Expense – the high-speed magnetic ink character readers are very expensive.

ATMs (automated teller machines) (cash points)

ATMs, commonly called cash points, are the "hole in the wall" cash dispensers that many people use when the bank is not open or when they do not want to queue inside the branch. In order for you to use the service the machine needs to check that you are the card holder. You are asked to enter a PIN (personal identification number) that only you should know. If the card is stolen then the thief should have no way of finding this, unless of course you have stupidly written it down!

Here are some of the things you can do using an ATM:

- You can get cash out.
- You can find out the balance in your account.
- You can change your PIN (personal identification number).
- You can make deposits (i.e., put cash, cheques or both into your account).
- You can obtain a mini statement listing your recent transactions (i.e., money in and out of your account).
- Other facilities such as topping up your phone card might be available.

Type of processing used with ATMs

Real-time transaction processing is used with ATMs. This means that as soon as a customer gets the money out of their account, their balance is updated.

An ATM, commonly called a cash point.

Processes involved in using an ATM

The card is inserted – the ATM connects to the bank.

PIN is entered – if incorrect it stops and allows a second try or checks to see if the card is stolen.

- if correct the ATM checks if the card is valid/not out of date.

If PIN is OK the system moves onto the next step, otherwise it stops.

Customer selects option.

- ATM checks balance.
- ATM checks if customer is trying to withdraw more than their daily allowance.

If OK the system asks if receipt is required.

If Yes – receipt is printed

- card is returned
- money is given out.

Note that the card is given out first and the money is given out last.

Benefits to banks in using ATMs:

There are some benefits to the banks in the use of ATMs and these include:

- Staff are freed from performing routine transactions so that more profitable sales-oriented work can be done.
- Fewer staff are needed, since the computer does much of the routine work.
- A 24-hour per day service is provided to satisfy their customers' demands.
- The system makes it impossible for a customer to withdraw funds from their account unless they have the money in their account or an agreed overdraft.
- Unusual spending patterns or location of withdrawal can trigger an alert to system not to issue cash without contacting the bank for authorization using a series of security questions.

Benefits to customers in using ATMs:

- Some customers prefer the anonymous nature of the machine since it cannot think you have stolen the cheque book or think that you are spending too much.
- It is possible to use the service 24 hours per day; ideal for those people who work irregular hours.
- It is possible to park near the dispenser of an evening, so getting cash is a lot quicker.
- Fewer queues, since the transactions performed by the ATM are a lot faster.
- If card is stolen, the thief cannot get money from the ATM unless they know your PIN.

Online banking

Many of the tasks you would have had gone to a bank branch to do, you can now perform at home using Internet or online banking. Using online banking you can:

- view bank statements
- transfer money between accounts
- make payments for bills
- apply for loans.

Online banking uses the Internet to enable a customer at home to connect to the bank ICT systems and interact with them. In order to do this the customer has to enter log-in details (username and password) and answer some other security questions.

When you access a secure link, the web address should start with https not just http.

Any details passed between the bank and their customers are encrypted to ensure hackers cannot access banking details.

Advantages to the bank of online banking:

▸▸ The bank can reduce the number of branches, which will reduce costs.
▸▸ Fewer bank workers are needed and this means lower total wage bill.
▸▸ The bank staff can be less qualified so this reduces the wage costs.

Disadvantages to the bank of online banking:

▸▸ Customers may feel lack of personal contact and this may cause some customers to move their account to a different bank.
▸▸ It is easier to sell products such as loans face-to-face and this can reduce bank profits.
▸▸ Have to spend out large amounts of money on new systems to perform the online banking.
▸▸ Have to pay out large amount of money on redundancies when staff lose their jobs.
▸▸ Have to employ and train staff with skills in computer/network development and maintenance.

Advantage to the customer of online banking:

▸▸ Do not need to spend time travelling to the bank to perform some banking transactions.
▸▸ Can perform banking transactions any hour of the day or night.
▸▸ Can pay bills without the need for cheques and money for postage/parking.
▸▸ Easier to check on your account balance so less likely to be charged for overdrafts, etc.

Disadvantages to the customer of online banking:

▸▸ The worry of hackers accessing your account and stealing money.
▸▸ You cannot withdraw cash so will still need to visit a branch or ATM.
▸▸ Lack of the personal touch you get when banking at a traditional branch.

QUESTIONS C

1 a A school uses a school management system. Describe **two** features of a school management system. *(2 marks)*
  b Explain **one** advantage in the school using a school management system *(1 mark)*
  c Explain **one** disadvantage in the school using a school management system. *(1 mark)*

2 ICT-based student registration systems in schools offer the advantages of speed and accuracy to the form teacher. Discuss **two** other advantages to the school in using ICT to register students. *(4 marks)*

3 Plastic cards, such as the one shown, are often issued to bank customers.

  a Give the names of **two** different types of plastic card that are used by bank customers. *(2 marks)*
  b These plastic cards often contain data in a magnetic stripe on the card but many of the new cards use chip and PIN.
    Explain why these new chip and PIN cards were introduced and what advantages they offer over the older cards. *(4 marks)*

4 One problem in using credit cards is that they can be used fraudulently.
  a Explain **two** ways that a credit card could be used fraudulently. *(2 marks)*
  b State **one** way such frauds can be prevented. *(1 mark)*

Applications in medicine

ICT is used extensively in medicine in the following ways:

- Medical databases – patient records are stored on a database and can be accessed from different places. There is simultaneous access to these records and they are easier to read than handwritten records.
- Patient identification using bar codes – bar codes are on patient wristbands so these can be scanned and the details obtained at the bedside.
- Hospital intranets – only hospital staff are allowed access to this network, which holds patient records. The intranet is also used internally for sending email between staff.
- Patient monitoring – sensors are used to measure vital signs such as temperature, blood pressure, pulse, central venous pressure, blood sugar and brain activity. Patient monitoring takes measurements automatically and frees up medical staff. There are no mistakes in the readings and the readings are taken in real time.
- Medical diagnosis – MRI (magnetic resonance imaging) and CAT (computed axial tomography) scanning equipment used computer modelling to view the inside of a patient without the need for operations.
- Expert systems – these are used to help less experienced doctors to make a more accurate diagnosis. These are covered later in this chapter.
- Computerized reporting of laboratory tests.
- Digital X-rays can be viewed faster than conventional X-ray film.
- Pharmacy records – generates labels for prescribed medicines, recording patient prescriptions, alerting pharmacist/doctors to allergies or interactions between drugs.

Applications in libraries

Most libraries are computerized and the systems usually make use of bar codes and a relational database.

Each member is given a unique member number, which acts as the primary key in the Members table of the database. Rather than type this number, the number is coded in the form of a bar code. This is faster and more accurate than typing it in.

Books are also bar coded with a unique number and when they are borrowed the member's ticket bar code is scanned along with the bar codes of all the books that have been borrowed. On their return the books are scanned thus telling the computer that the borrower has returned the books. The system can identify overdue books and it is possible to use mail merge to send letters, or email members asking them to return the books.

The book table can be used to locate a specific book and to find out whether it is on the shelf or has been borrowed. It is possible to automatically generate lists of books (e.g. by author). Books can be reserved using the system so when the book arrives back, it is kept aside rather than being put back on the shelf.

Expert systems

An expert system is an ICT system that uses artificial intelligence to make decisions based on data supplied in the form of answers to questions. This means that the system is able to respond in the way that a human expert in the field would to come to a conclusion. A good expert system is one that can match the performance of a human expert in the field.

Expert systems consist of the following components:

- Knowledge base – a huge organized set of knowledge about a particular subject. It contains facts and also judgemental knowledge, which gives it the ability to make a good guess, like a human expert.
- Inference engine – is a rules base and is the part of the expert system that does the reasoning by manipulating and using the knowledge in the knowledge base. There is usually a way of phrasing a question to the system or a way of searching for information using a search engine.
- User interface – this uses an interactive screen (which can be a touch screen) to present questions and information to the operator and also receives answers from the operator.

The processes involved in a typical expert system such as a medical expert system for diagnosis are as follows:

1 An interactive screen appears and asks the user questions.
2 Answers are typed in or options are selected on a touch screen.
3 The inference engine matches the data input with the knowledge base, using the rules base until matches are found.
4 The system suggests the probable diagnosis and suggests treatments.

Expert systems/IKBS are computer systems that behave like a human expert on a subject.

Applications of expert systems

Expert systems can be used for all sorts of applications and here are some of them.

Medical diagnosis

A health service website enables users to determine whether they need to call a doctor or visit a hospital. An expert system guides them through a series of questions, which they answer by clicking on a choice of answers. Such expert systems can determine whether someone with pains is having a heart attack or simply has indigestion.

Prospecting for minerals and oil

Using geological information, an expert system can use the information to determine the most likely places to choose for further exploration. This reduces the cost of mineral or oil exploration.

Expert systems can enable doctors to make an expert diagnosis of a patient's illness.

Expert systems can use geological data to predict where best to strike oil.

For giving tax advice to individuals and companies

Tax is complex and a lot of expertise is needed in order to give the correct advice. This is where expert systems come in. They are able to store a huge amount of data and they can ask the user a series of questions and come up with expert advice on how to pay less tax.

Car engine fault diagnosis

Modern car engines are very complex and when they go wrong it is hard for engineers to know what the problem is. Using an expert system created by the car manufacturer the engineers can be guided through a series of tests until the exact fault is identified.

Chess games

Chess game software is an expert system because it mimics an expert human chess player.

Advantages of expert systems:

▸▸ Fewer mistakes – human experts may forget but expert systems don't.

▸▸ Less time to train – it is easy to copy an expert system but it takes many years to train a human expert.

▸▸ Cheaper – it is cheaper to use an expert system rather than a human expert because human experts demand high wages.

▸▸ More expertise than a single expert – many experts can be used to create the data and the rules, so the expert system is a result of not one but many experts.

▸▸ Always asks a question a human expert may forget to ask.

Disadvantages of expert systems:

▸▸ Lack common sense – humans have common sense, so they are able to decide whether an answer is sensible or ridiculous. Human experts can make judgements based on their life experiences, and not just on a limited set of rules as is the case with computer systems.

▸▸ Lack senses – the expert system can only react to information entered by the user. Human experts have many senses that they can use to make judgements. For example, a person describing a type of pain might use body language as well, which would not be detected by an expert system.

▸▸ The system relies on the rules being correct – mistakes could be made that make the system inaccurate.

QUESTIONS D

1 ICT is used in hospitals for patient monitoring.
   a Sensors are used to monitor a patient. Give the names of **three** sensors that would be used. *(3 marks)*
   b Other than patient monitoring give **two** uses of ICT in a hospital. *(2 marks)*

2 Expert systems are used in medicine to help diagnose illnesses.
   a An expert system consists of three components. Name **two** of these components. *(2 marks)*
   b Give **one** benefit to the patient in using an expert system. *(1 mark)*
   c Give **one** benefit to the doctor in using an expert system. *(1 mark)*
   d Give **one** possible disadvantage in using this type of expert system. *(1 mark)*

3 Doctors and hospital consultants often make use of expert systems in their work.
   a Explain what is meant by an expert system. *(2 marks)*
   b Describe **one** way in which a doctor or hospital consultant can make use of an expert system. *(2 marks)*

4 Expert systems are becoming popular uses for ICT.

   a Name **two** jobs that are likely to use an expert system. *(2 marks)*

   b For each of the jobs you have named in part (a) explain how the expert system would be used. *(2 marks)*

5 One example of where an expert system is used is to help doctors diagnose illnesses. Give **one** other use of an expert system. *(2 marks)*

6 Expert systems have many uses. For example, one expert system called MYCIN is used by doctors to pinpoint the correct organism in blood that is responsible for a blood infection.

   a By referring to the above example give **one** advantage of using an expert system rather than a doctor. *(1 mark)*

   b By referring to the above example, give **one** disadvantage in using the expert system rather than a doctor. *(1 mark)*

Applications in the retail industry

Point of sale (POS) and electronic funds transfer at point of sale (EFTPOS) systems

Point of sale terminals are the computerized tills where you take your goods for payment in a shop. Many of these terminals also allow payment from your bank account to the store's bank account directly using a debit card. These terminals are called EFTPOS systems. EFTPOS systems can also allow the customer to get "cashback" which moves money from their bank account to the store and this money is given to the customer as cash.

Because the terminals are networked together, when an item is sold and its bar code is scanned, the system looks up the price and description details to print out an itemized receipt. At the same time the system will deduct the item from stock so that the stock control system is updated.

POS/EFTPOS terminals consist of the following hardware:

▸▸ Bar code reader/laser scanner – this is used to input a number that is coded in the bar code as a series of light and dark lines.

▸▸ Keyboard – the keyboard is used to enter codes on items if the bar code is damaged.

▸▸ Touch screens – these are often used in restaurants where there are no goods to scan.

▸▸ Swipe card readers – these are used to swipe the magnetic stripes on loyalty cards.

▸▸ Chip and PIN readers – these are used by customers to insert their credit/debit cards containing a chip. The system then asks them to enter their PIN (personal identification number) which is a number only they know. This is compared with the number encrypted and stored on the card. If the number

entered matches that stored, it proves to the system that they are the genuine owner of the card.

Point of sale terminals are connected via networks to other systems such as:

▸▸ Loyalty card systems – where customers are given loyalty points according to how much they spend.

▸▸ Accounts systems – where the money coming into the shop is accounted for.

▸▸ Automatic stock control systems – the system knows what has been sold, so that it can automatically reorder more once the stock falls below a certain level.

This reader can read chip and PIN as well as swipe cards.

Automatic stock control

When an item is sold at the POS/EFTPOS terminal the number of that particular item in stock is reduced by one. This means that the computer knows how many items are in stock. Once the number of items has fallen below a certain level, the computer system will automatically order more stock from the supplier. This means that stores should not run out of fast-selling items. Good stock control systems are very important in supermarkets as customers will go elsewhere if the shop keeps running out of key items such as bread, milk, etc.

Touch screens are used as input devices in restaurants and bars.

Internet shopping

Internet shopping means purchasing goods and services using the Internet.

Most businesses have web sites to show the products and services available. Lots of these web sites allow customers to browse online catalogues and add goods to their virtual shopping basket/trolley just like in a real store. When they have selected the goods, they go to the checkout where they have to decide on the payment method. They also have to enter some details such as their name and address and other contact details. The payment is authorized and the ordering process is completed. All that is left is for the customer to wait for delivery of their goods.

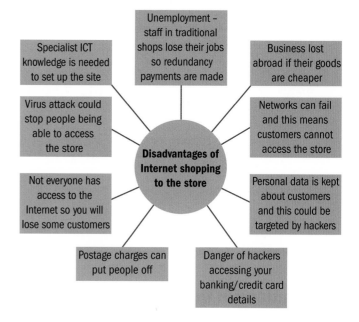

Advantages of Internet shopping to customers
- No opening hours – goods/services can be bought 24/7
- Goods/services are cheaper because of lower costs of Internet business
- Goods are delivered to your home – ideal if people cannot get out because they are elderly or disabled
- Wider range of goods to choose from
- Cost savings are passed to customers with cheaper goods
- No travelling costs to go shopping
- Worldwide marketplace – you can order goods from anywhere in the world

Disadvantages of Internet shopping to customers
- Shoppers may be worried about entering their credit/debit card details
- Sometimes more hassle when returning goods, e.g. wrong size, faulty, etc.
- Harder to assess the quality of the goods before ordering
- Lack of personal service you get in a store
- Loss of the social pleasure of going shopping
- Hidden costs of postage and packing
- Problems with bogus sites where you pay for goods that never arrive

Advantages of Internet shopping to the store
- Can sell goods 24/7 – no opening and closing times
- No expensively fitted out high street stores to pay for
- Because of high volume of goods sold they can negotiate good discounts with suppliers
- Do not need to work long hours like in a shop
- Less shoplifting
- Can sell goods to anyone in the world
- Can site warehouse anywhere that rental is cheap
- Cheaper to keep in touch with customers as they can be emailed

Disadvantages of Internet shopping to the store
- Specialist ICT knowledge is needed to set up the site
- Unemployment – staff in traditional shops lose their jobs so redundancy payments are made
- Business lost abroad if their goods are cheaper
- Virus attack could stop people being able to access the store
- Networks can fail and this means customers cannot access the store
- Not everyone has access to the Internet so you will lose some customers
- Personal data is kept about customers and this could be targeted by hackers
- Postage charges can put people off
- Danger of hackers accessing your banking/credit card details

Test yourself

The following notes summarize this topic. The notes are incomplete because they have words missing. Using the words in the list below, copy out and complete the sentences A to M, underlining the words that you have inserted. Each word may be used more than once.

inference engine   knowledge base   online
cheaper   diagnosis   batch   experts   sensors
user interface   expert   real-time   question

A _____ processing is where all the inputs are put together and then processed in one go.

B The type of processing appropriate for processing cheques during cheque clearing is _____ processing.

C The system that is automatically updated when a change (called a transaction) is made uses _____ processing.

D Traffic light systems and flood control systems always use _____ processing.

E Computer control systems use _____ as the input devices.

F An _____ system is an ICT system that uses artificial intelligence to make decisions based on data supplied in the form of answers to questions.

G A _____ is a huge organized set of knowledge about a particular subject used by an expert system.

H The set of rules on which to base decisions on used by expert systems is called the _____.

I The part of the expert system where the use interacts with the system is called the _____.

J Expert systems have the advantage that they are usually built using the knowledge of many _____.

K Expert systems have the advantage that they will not forget to ask a _____ nor will they forget things.

L One application of expert systems is medical _____ where the expert system is used by a less experienced doctor to help make an expert and accurate diagnosis.

M Expert systems have the advantage in that it is much _____ than consulting one or several human experts.

REVISION QUESTIONS

1 **a** Give the name of a household device that uses a control system. [1]

   b Explain how the control system controls the device you have named in part (a). [3]

2 A school is using a traditional paper-based registration system.
The school uses registers to record morning and afternoon attendance details for each student.

   a Describe **three** advantages in using ICT systems for registration. [3]

   b Describe **three** disadvantages in using ICT systems for registration. [3]

3 Give **two** examples of tasks that are completed by robots. [2]

4 Online banking is very popular with home users of ICT.

   a Name and describe **three** services offered by online banking. [3]

   b Some people are sceptical about online banking. Describe **two** worries that people might have with online backing. [2]

   c Describe **one** way that the banks can address one of the worries you have described in part (b). [1]

5 Many people use the Internet to access booking systems. By referring to a relevant example, explain how an Internet booking system works and the advantages to the home user in being able to book tickets/seats online. [5]

EXAM AND EXAM-STYLE QUESTIONS

1 A floor turtle can use the following instructions:

INSTRUCTION	MEANING
FORWARD n	Move n mm forward
BACKWARD n	Move n mm backward
LEFT t	Turn left t degrees
RIGHT t	Turn right t degrees
PENUP	Lift the pen
PENDOWN	Lower the pen
REPEAT n	Repeat the following instructions n times
END REPEAT	Finish the REPEAT loop

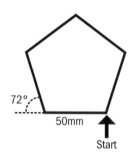

72°

50mm

Start

Complete the set of instructions to draw this shape by filling in the blank lines. *[5]*

PENDOWN

.... 90

REPEAT ....

FORWARD ....

.... 72

....

(Cambridge IGCSE Information and Communication Technology 0417/13 q6 Oct/Nov 2010)

2 An oil company is investigating whether they are likely to find oil at a certain site. They will use an expert system to help them. There are a number of inputs and outputs used with such a system. List **four** examples of these. *[4]*

(Cambridge IGCSE Information and Communication Technology 0417/13 q12 Oct/Nov 2010)

3 A car repair centre uses an expert system to help diagnose car engine faults.

   a Describe the inputs, outputs and processing of this system. *[6]*

   b Give **two** other examples of situations where expert systems might be used. *[2]*

(Cambridge IGCSE Information and Communication Technology 0417/01 q11 Oct/Nov 2009)

4 **a** Give **three** applications which use online processing. *[3]*

   b Give **three** applications which use batch processing. *[3]*

8 Systems analysis and design

The systems life cycle is the series of stages that are completed when developing a new system or improving an old one. The stages are carried out in order and this ensures that the system is developed properly. The existing system might be paper-based or, more probably, uses computers but is no longer good enough or up-to-date. In this chapter you will learn about systems analysis and the tasks that are completed for each stage.

The key concepts covered in this chapter are:
▸▸ Understand the purpose of systems analysis and design
▸▸ Understand what is involved in the analysis stage
▸▸ Understand what is involved in the design stage
▸▸ Understand what is involved in the development and testing stage
▸▸ Understand what is involved in the implementation stage
▸▸ Understand what is involved in the documentation stage
▸▸ Understand what is involved in the evaluation stage

The purpose of systems analysis and design

It is important that new systems being created are fit for purpose. Systems developed without too much thought tend to have serious faults. To prevent this happening, most systems are developed in a series of set stages completed in a set order. These are called the stages of systems analysis and design.

These stages are as follows:

▸▸ Analysis
▸▸ Design
▸▸ Development and testing
▸▸ Implementation
▸▸ Documentation
▸▸ Evaluation.

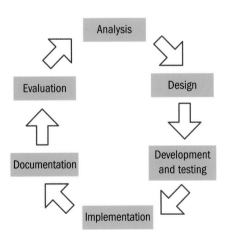

Analysis

Analysis looks in detail at the current system or the requirements for a task that has never been performed before.

Analysis will normally involve the following:

▸▸ Identifying the problem(s) that needs solving.
▸▸ Collecting facts about the old system or the required system using questionnaires, interviews, etc.
▸▸ Identifying the inputs, outputs, and processing of the current system or the new system if an existing system does not exist.
▸▸ Identifying problems with the existing system.
▸▸ Identifying the user and information requirements necessary to solve the problem(s).

The different methods of researching a situation

The starting point of analysis involves finding out what people want from the new proposed system or looking at the existing system to find out how it works and might be improved. There are several ways this can be done:

▸▸ using questionnaires
▸▸ using interviews
▸▸ using observation
▸▸ examination of existing documentation.

Using questionnaires

A questionnaire could be given to each user and left with them for completion. Questions should be about how the job is done now and not about the overall running of the business. It could also be about the information the new system needs to give them.

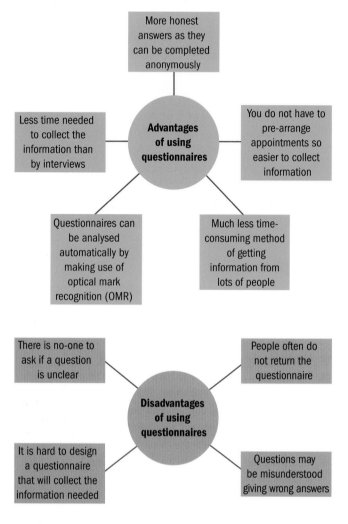

Using interviews

Interviews take longer than questionnaires, so this method is good if there are only a few users of the system. People at the different levels in the organization who will use the new system should be interviewed. At these interviews you can find out how the existing system works and what things are required from the new system.

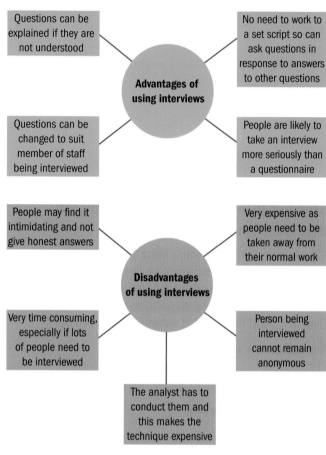

Using observation

Here you sit with someone who is actually doing the job the new system is designed to do. You then see the problems encountered with the old system as well as chat to the user about what the new system must be able to do.

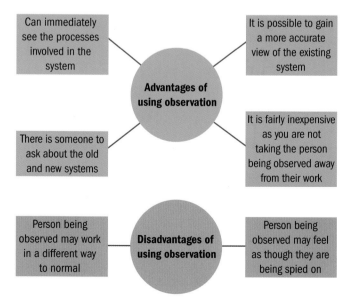

Questionnaires can be used to collect information about the new system from lots of people.

Examination of existing documentation

This involves looking at any of the paperwork involved with the current system. This would include documents such as order forms, application forms, lists of stock, and so on. You can also look at the records that are kept in filing cabinets.

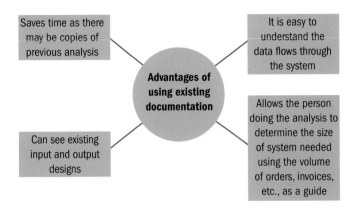

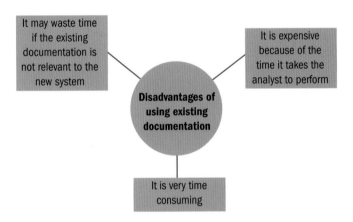

Design

After the analysis stage, the requirements for the new system should be thoroughly understood and work can start on how best to design the new system.

You would never start building a house without proper designs and plans. ICT systems are no different, as it is much easier to design systems carefully rather than have to change them at a later stage. ICT systems can be very different but it is usually necessary to design documents, files, forms/inputs, reports/outputs and validation methods.

Producing the designs

All ICT systems are different. For example, you might be creating a website or a database. Each must be designed carefully by completing the relevant design tasks, which would include:

▶▶ Designing data capture forms (these are forms used for the input of data).
▶▶ Designing screen layouts (these are part of the user interface).
▶▶ Designing validation routines that help prevent invalid data from being processed.
▶▶ Deciding on the best form of verification for the system.

▶▶ Designing report layouts (this is the output from the system that is printed).
▶▶ Designing screen displays for the output (this is the output from the system that is displayed on the screen).
▶▶ Designing the required data/file structures (for example, if a database needs to be produced then the tables will need to be designed).

Designing validation routines

Validation is a check performed by a computer program during data entry. Validation is the process that ensures that data accepted for processing is sensible and reasonable. For example, a living person's date of birth could not be before 1892 as in 2012 this would make them 120 years old (the current oldest person is 115). Validation is performed by the computer program being used and consists of a series of checks called validation checks.

When a developer develops a solution to an ICT problem, they must create checks to reduce the likelihood of incorrect data being processed by the computer. This is done by restricting the user as to what they can enter, or checking that the data obeys certain rules.

Validation checks

Validation checks are used to restrict the user as to the data they can enter. There are many different validation checks, each with their own special use including:

▶▶ Boolean checks – data is either: True or False, Y or N.
▶▶ Data type checks – these check that data being entered is the same type as the data type specified for the field. This would check to make sure that only numbers are entered into fields specified as numeric or letters only in a name.
▶▶ Presence checks – some database fields have to be filled in, whilst others can be left empty. A presence check would check to make sure that data had been entered into a field. Unless the user fills in data for these fields, the data will not be processed.
▶▶ Length checks – there is a certain number of characters that need to be entered. For example, in one country a driving licence number has a length of 15 characters. Without the correct number of characters, the data would be rejected.
▶▶ Consistency checks – checks to see if the data in one field is consistent with the data in another field. For example, if gender is M then there should not be Miss in the title field.
▶▶ Range checks – are performed on numbers. They check that a number being entered is within a certain range. For example, all the students in a college are aged over 14, so a date of birth being entered which would give an age less than this would not be allowed by the range check.
▶▶ Format checks – are performed on codes to make sure that they conform to the correct combinations of characters. For example, a code for car parts may consist of three numbers followed by a single letter. This can be specified for a field to restrict entered data to this format, e.g. DD/MM/YY for a date field.

▸▸ Check digits – are added to important numbers such as bank account numbers, International Standard Book Numbers (ISBNs), etc. Check digits are placed at the end of the block of digits and are used to check that the digits have been entered correctly into the computer. When the large number is entered, the computer performs a calculation using all the digits to work out this the check digit. If the calculation reveals that the check digit is the same as that calculated by the other digits, it means that the whole number has been entered correctly.

Choosing a method of verification

Verification means checking that the data being entered into the computer perfectly matches the source of the data. For example, if details from an order form were being typed in using a keyboard, then when the user has finished, the data on the form on the screen should be identical to that on the paper form (i.e., the data source). Also if data was sent over a network, the data needs to be checked when it arrives to make sure no errors are introduced during transmission.

Here are some methods of verification:

▸▸ Visual comparison of the data entered with the source of the data – involves one user carefully reading what they have typed in and comparing it with what is on the data source (order forms, application forms, invoices, etc.) for any errors, which can then be corrected.

▸▸ Double entry of data – involves using the same data source to enter the details into the computer twice and only if the two sets of data are identical will they be accepted for processing. The disadvantage of this is that the cost of data entry is doubled. Double entry of data is often used when creating accounts over the Internet. They may ask you to create a password and enter it twice. This ensures there are no mistakes that would prevent you from accessing the account.

⊙ KEY WORDS

Check digit a decimal number (or alphanumeric character) added to the end of a large number such as an account number for the purpose of detecting the sorts of errors made on data entry. All the digits are used in a calculation which calculates the check digit. This calculated number is then compared with the check digit to see if the two are the same. If so, then the large number has been input correctly.

Range check data validation technique that checks that the data input to a computer is within a certain range. If the data is outside the range, then it is rejected.

Validation checks checks a developer of a solution sets/ creates, using the software, in order to restrict the data that a user can enter, so as to reduce data entry errors. Validation checks check that the data being entered is reasonable.

Verification checking that the data being entered into the ICT system perfectly matches the source of the data. Verification can involve proof-reading or the entry of data twice.

QUESTIONS A

1 **a** What is meant by a check digit? *(3 marks)*
      b Give **two** different examples where check digits are used. *(2 marks)*

2 Here are some dates of birth that are to be entered into an ICT system:
      a 12/01/3010
      b 01/13/2000
      c 30/02/1999
      Assume that all the dates are in the British format dd/mm/yyyy. For each one, explain why they cannot be valid dates of birth. *(3 marks)*

3 When an employee joins a company they are given an employee code.
      a Here is an example of an employee code:
             LLLNNNNNN where L is a letter of the alphabet and N is a number.
             Describe **one** type of validation that could be used with this field. *(2 marks)*
      b Employees are given an annual salary. Describe **one** type of validation that could be used with this field. *(2 marks)*

4 A computer manager says, "data can be valid yet still be incorrect". By giving **one** suitable example, explain what this statement means *(3 marks)*

5 An online form for ordering DVDs uses a presence check for some of the fields.
      a Describe what a presence check is and why some fields have them while others don't. *(3 marks)*
      b Give **one** field that might have a presence check and **one** field that would not need a presence check. *(2 marks)*

6 During the development of a new system it is important to find out how the existing system works if there is one. The usual way to do this is to carry out a fact search/fact find.

      Name and describe **three** different ways of collecting facts about an existing system. *(6 marks)*

Development and testing

Development of the new system means creating the new system following the designs created in the previous stage. The new system will need to be thoroughly tested during and after the development.

Developing the system from the designs and testing it

Creating the data/file structure and testing it

In the case of a database this would involve creating the database tables and entering the field names, the type of each field, the length of the field, etc., and setting up key fields. The links between the tables will need to be created. Test data can then be entered for each field to check that the data can be extracted in the way required.

Creating validation routines and testing them

The validation routines/checks will have been decided on in the design stage. These designs will be used with the software to create the checks that the data being entered is reasonable or acceptable. Each validation check will need to be thoroughly tested by entering some test data, which will be discussed later.

Creating input methods and testing them

There are many different ways of inputting data into a system. For example, share prices can be input directly from a website and put into a spreadsheet. Many systems will make use of a keyboard to enter data into a form, spreadsheet, etc. As well as creating these methods, they will need to be tested.

Creating output formats and testing them

This involves designing of output such as reports from databases, output on screen, etc. Testing of the output involves checking that the information output is complete and that it produces the correct results.

Testing strategies

Each module of the ICT solution created needs to be thoroughly tested to make sure it works correctly. The modules are tested with data that has been used with the existing system so that the results are known.

Testing strategies often make use of a test plan. A test plan is a detailed list of the tests that are to be conducted on the system, when it has been developed, to check it is working properly. These test plans should be comprehensive. A good way of making sure that they are is to make sure that:

▸▸ tests are numbered
▸▸ each test has the data to be used in the test clearly specified
▸▸ the reason for the test is stated
▸▸ the expected result is stated.

Space should be left for:

▸▸ the actual result and/or comment on the result
▸▸ a page number reference to where the hard copy evidence can be found.

Testing using normal, abnormal, and extreme data

Testing should always be performed with the following three types of data:

▸▸ Normal data – is data that should pass the validation checks and be accepted.
▸▸ Abnormal data – is data that is unacceptable and that should be rejected by the validation check. If data is rejected then an error message will need to be displayed explaining why it is being rejected.
▸▸ Extreme data – is data on the borderline of what the system will accept. For example, if a range check specifies that a number from one to five is entered (including one and five) then extreme data used would be the numbers one and five.

Here is a test plan to test a spreadsheet for analysing the marks in an examination. A mark is input next to each candidate's name. The mark is a percentage and can be in the range 0% to 100%. In this exam, half marks are possible.

Test no	Test mark entered	Purpose of test	Expected result	Actual result
1	45	Test typical data	Accept	
2	0	Test extreme data	Accept	
3	100	Test extreme data	Accept	
4	123	Test abnormal data	Reject and error message	
5	-3	Test abnormal data	Reject and error message	
6	D	Test abnormal data	Reject and error message	

The "actual result" column would be filled in when the test mark was entered. If the expected result and the actual results are all the same then the validation checks are doing their job. If the two do not agree, then the validation checks will need to be modified and re-tested.

Testing the whole system

Once all the modules have been tested separately and any problems with them solved, the modules can be joined to create the whole system. The whole system is then thoroughly tested and any problems addressed.

Implementation

In order to change from one system to another it is necessary to have a way of doing this and this is called a method of system implementation. There are several methods used and these are outlined here.

Methods of implementation

Direct changeover

With direct changeover you simply stop using the old system one day and start using the new system the next day.

Advantages of direct changeover include:

▸▸ Fastest method of implementation.
▸▸ Benefits are available immediately.
▸▸ Only have to pay one set of workers who are working on the new system.

Disadvantages of direct changeover include:

▸▸ If the new system fails, you might lose all the data.
▸▸ All staff need to be fully trained before the change, which may be hard to time/plan.
▸▸ The old system is removed so there is no system to go back to if things go wrong.

Parallel running

This method is used to minimize the risk in introducing a new computer system. Basically, the old computer system is run alongside the new computer system for a period of time until all the people involved with the new system are happy it is working correctly. The old system is then switched off and all the work is done entirely on the new system.

Advantages of parallel running include:

▸▸ If the new system fails then no data will be lost.
▸▸ It allows time for the staff to be trained gradually.

Disadvantages of parallel running include:

▸▸ It is expensive in terms of people's time.

Phased implementation

A module at a time can be converted to the new system in phases until the whole system is transferred.

Advantages of phased implementation include:

▸▸ Only need to pay for the work to be done once.
▸▸ Training can be gradual as staff need only to train in the module required each time.
▸▸ If the new system fails, most of the company/organization can still use the old system.
▸▸ If the new system fails, you only need to go back to the latest phase and do not have to review the whole system.
▸▸ IT staff can deal with problems caused by a module before moving on to new modules.

Disadvantages of phased implementation:

▸▸ If there is a problem, then some data may be lost.
▸▸ There is a cost in evaluating each phase before implementing the next.
▸▸ It is only suitable for systems consisting of separate modules.
▸▸ It can take a long time before the whole system is implemented.

Pilot running

This method is ideal for large organizations that have lots of locations or branches where the new system can be used by one branch and then transferred to other branches over time.

Advantages of pilot running:

▸▸ The implementation is on a much smaller and more manageable scale.
▸▸ There is plenty of time available to train staff.

Disadvantages of pilot running:

▸▸ It can take a long time to implement the system across the whole organization.
▸▸ If the system fails in one of the branches, data can be lost.

Which method should be chosen?

▸▸ For organizations or departments within organizations which need a quick changeover, direct changeover is best. However, this method is risky as some data could be lost if the system malfunctions but it is also relatively inexpensive because the work is not having to be done twice as with parallel running.
▸▸ For organizations or departments within organizations which cannot afford to lose data, parallel running, phased implementation or pilot running can be used. Parallel running offers less risk because the old system can still be used if there are problems but takes much longer and is more expensive because everything is done twice.

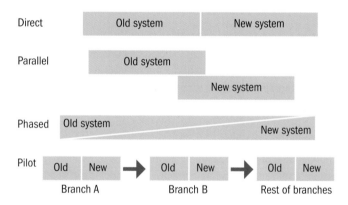

Documentation

Once a new system has been produced it needs to be documented. Two sets of documentation are produced with one set aimed at the user and the other set at the developers who may need to alter the system in the future.

User documentation

User documentation is documentation that the user can turn to for learning a new procedure or for dealing with a problem that has cropped up. User documentation will normally consist of the following:

▸▸ the hardware requirements to run the system
▸▸ the operating system needed to run the software

- how to run the program/use the system
- how to log in and log out of the system
- how to perform tasks such as enter data, sort data, search for data, save data, produce printouts, etc.
- details of sample runs
- tutorials to help a user become familiar in using the system
- details of input and output formats (e.g. screen layouts and print layouts)
- error messages and how to deal with them
- trouble-shooting guide
- frequently asked questions/ FAQ/help guide.

Technical documentation

Technical documentation is aimed at people who may maintain the new system or develop systems in the future. Technical documentation will normally consist of the following:

- purpose of the system
- hardware requirements and software requirements
- a copy of the system design
- copies of all the diagrams used to represent the system (program flowcharts, system flowcharts, network diagrams, etc.)
- program listings/program coding
- lists of variables used
- details of known bugs
- sample runs (with test data and results)
- file structures (e.g., structure of database tables, etc.)
- validation routines used
- user interface designs
- test plans
- meaning of error messages.

Evaluation

Evaluation takes place soon after implementation. It is only then that the users and others involved in the development of the system will find out about any problems with the new system.

The need for evaluation

Evaluation of a new system will look at each of the following:

- The efficiency of the solution – this looks at how well the system works, such as is it fast, does it give people the information they require, and so on.
- The ease of use – is the system easy to use? Has a good user interface been developed? Are training costs minimized because it is so easy to use?
- The appropriateness of the solution – this can involve checking that the original user requirements have been fully met by the new system and assessing how happy the clients are with the development of the new system.

Evaluation strategies

Ways of evaluating a system include:

- Comparing the system developed with the initial requirements.
- Evaluating the users' responses using the system.
- Comparing the performance of the new system with that of the old system.
- Identifying any limitations that need to be made to the system.
- Identifying any improvements that need to be made to the system.
- Interviewing users to gain an insight into problems with the system.
- Seeing if cost efficiencies have been met (e.g., a person getting through more work than with the old system, etc.).

Once they are developed, systems may still need to be changed in some way. These changes can be changes in hardware or software. Here are some of the changes that may need to be made:

- Changes in the way the business or organization operates may require that the system be altered. For example, a change in the rate of sales tax or changes in income tax could mean that software will need altering.
- Users of the new system may ask for extra functions to be added. The existing software will need to be changed to do this.
- Users may report system crashes which need to be investigated and are often solved by changes to the software.
- Poor performance or bugs in the software will be reported during evaluation meetings so they can be corrected.
- Slow performance may be reported and corrected by changes to hardware.

QUESTIONS B

1 Here is a list of the steps that are stages of the systems life cycle. At present these are in the wrong order. Put the steps in the correct order. *(5 marks)*

   Documentation
   Evaluation
   Analysis
   Development and testing
   Implementation
   Design

2 Here are some of the steps that are stages in the systems life cycle.

   Development and testing
   Implementation
   Design
   Analysis
   Evaluation

   Write down the name of a step from the list above where the following tasks would be carried out:

a Planning the construction of the new system. *(1 mark)*

b Planning the testing of the new system. *(1 mark)*

c Getting the user to answer a questionnaire to find out what is required from the new system. *(1 mark)*

d Asking users what they think of the new system that has been developed. *(1 mark)*

e Putting data into the computer to check if the output is what was expected. *(1 mark)*

3 When one ICT system is being replaced by another there are a number of different ways of doing this. One way is direct changeover.

a Describe the system strategy for implementing systems called direct changeover. *(2 marks)*

b Give **one** advantage and **one** disadvantage of direct changeover. *(2 marks)*

Activity 8.1

Produce a mind map (either hand drawn or produced using mind mapping software) that can be used to summarize what is involved in each of the system implementation strategies shown here. Your mind map should also show the advantages and disadvantages for each method:

➤ Direct changeover ➤ Parallel running

➤ Phased implementation ➤ Pilot running.

❗ Revision Tip

You may be asked to discuss the different implementation strategies for the organization mentioned in a question. Remember that you need to discuss this in relation to this organization and not a general discussion that could apply to any business. Also, remember that a "discuss" question needs to be answered in complete sentences. When you do this, try to ensure that each sentence contains one or more points.

Test yourself

The following notes summarize this topic. The notes are incomplete because they have words missing. Using the words in the list below, copy out and complete the sentences A to K, underlining the words that you have inserted. Each word may be used more than once.

> development and testing normal input
> evaluation documented training implementation
> whole direct changeover facts parallel running

A During the analysis stage, the person developing the system will collect _____ about the system using techniques such as questionnaires, observation and examination of existing documents.

B During the design stage the _____, processes, and output will be designed.

C With the design stage complete, the working version of the solution can be produced. This is called the _____ stage.

D Testing involves testing each module with _____, abnormal, and extreme data.

E Testing also involves testing the _____ system when all the modules are put together to form the whole system.

F _____ involves stopping using the old system and starting to use the new system.

G There are four main ways of implementing a system called _____ , parallel running, pilot running, and phased implementation.

H _____ involves the old ICT system being run alongside the new ICT system for a period of time until all the people involved with the new system are happy it is working correctly.

I Users need to know how to use the new system. The processing of helping them to understand the new system is called user _____.

J Once a new system has been developed it will need to be _____ and there are two types called user documentation and technical documentation.

K A review of the development of the project is completed at the end and this is called _____.

REVISION QUESTIONS

1 When students join the senior school, a form is filled in by their parents. The details on the form are then typed into a computer. The details are verified after typing. Explain briefly, how the details may be verified. [2]

2 A person's date of birth is entered into a database. State **three** things the validation program could check regarding this date as part of the validation. [3]

3 When a new member joins a fitness club they are given a membership number. The membership number is made up in the following way:

▸▸ Date of birth as six numbers.

▸▸ The final two figures of the year in which the customer joins the fitness club.

▸▸ A letter which is either J or S depending on whether they are a junior or senior member.

a Write down the membership number for a junior member who joined the club in 2010 and was born on 21/05/98. [1]

b When the membership number is entered into the database, it is validated. Describe what is meant by data validation. [2]

c Two examples of data validation are:

▸▸ range check

▸▸ format check.

Describe how these two methods could be used on the membership number field described above. [2]

4 State the names of **three** different methods of conversion from one ICT system to another and describe an advantage for each method. [6]

EXAMINATION QUESTIONS

1 Identify **three** methods which could be used to implement a new system. *[3]*

(Cambridge IGCSE Information and Communication Technology 0417/13 q17 Oct/Nov 2010)

2 When a new system is implemented, documentation is provided with it. Identify **four** items which would be found in technical documentation but **not** in user documentation.

[4]

(Cambridge IGCSE Information and Communication Technology 0417/13 q18 Oct/Nov 2010)

3 Joan owns a small company. She wishes to replace the existing computerised system with a new one. She has employed a systems analyst, Jasvir, to plan this.

a Before Jasvir decides on a system he must collect information about the existing system. Tick whether the following statements about the various methods of information collection are **TRUE** or **FALSE**.

	TRUE	FALSE
Examining documents has to be done in the presence of workers	☐	☐
Appointments have to be made in order to complete a questionnaire	☐	☐
It is possible to change questions in the course of an interview	☐	☐
Observing the current system can provide a detailed view of the workings of the system	☐	☐

[4]

b After Jasvir has completed the analysis of the existing system, he will need to design the new system. Tick **four** items which would need to be designed.

Inputs to the current system	☐
User and information requirements	☐
Data capture forms	☐
Validation routines	☐
Problems with the current system	☐
File structure	☐
Report layouts	☐
Limitations of the system	☐

[4]

Cambridge IGCSE Information and Communication Technology 0417/11 q14 May/June 2010

4 A systems analyst has been asked by a librarian to develop a computer system to store information about books and borrowers. After the existing system is analysed the new system will be designed. The first item to be designed will be the input screen.

a Name **four** items of data about **one** borrower, apart from the number of books borrowed, that would be input using this screen. *[4]*

b Describe **four** features of a well-designed input screen. *[4]*

c The librarian will need to type in data about each book from existing records. In order to prevent typing errors the data will be verified. Describe **two** methods of verification which could be used. *[4]*

d After the system is designed it will need to be implemented and then tested.
No borrower can take out more than 6 books.
Describe the **three** types of test data that can be used, using a number of books as an example for each. *[6]*

e The system must now be evaluated. Tick **three** reasons why this is done.

	✓
Improvements can be made.	
The hardware and software can be specified.	
Limitations of the system can be identified.	
To see how many books are required.	
To make sure the user is satisfied with the system.	
So that program coding can be written	

[3]

f After the system is implemented the librarian will be given technical documentation and user documentation. Name **three** different components of each type of documentation (Technical/User). *[6]*

(Cambridge IGCSE Information and Communication Technology 0417/01 q9 Oct/Nov 2009)

Communication

This is the first topic in the practical section and in it you will be using certain useful features of email which you may not have used before. You will also be finding out how to organize your emails so that they can easily be found in the future. Another important aspect of communicating information is to be able to find relevant information on the Internet quickly by the construction of searches. The activities here will instruct you in how to do this.

The key concepts covered in this chapter are:
▶▶ Sending and receiving documents and other files electronically
▶▶ Managing contact lists effectively
▶▶ Locating specified information from a given web site URL
▶▶ Finding specified information using a search engine
▶▶ Downloading and saving information as specified

Communicating with other ICT users using email

Communication by email has taken from over the use of letters as the way people in businesses communicate with each other.

What is an email?

An email is an electronic message sent from one communication device (computer, telephone, mobile phone, or PDA) to another. All web browser software has email facilities. There are many email facilities but those shown here are the main timesaving ones.

Search: Search allows you to find an email using keywords in the title or you can search for all the emails from or to a certain email address.

Reply: This allows you to read an email and then write the reply without having to enter the recipient's email address. As the recipient is sent both the original email and your reply they can save time because they know what your email is about.

Forward: If you are sent an email that you think others should see, you can forward it to them. An email, for example sent to you by your boss, could be forwarded to everyone who works with you in a team.

Address book: In the address book are the names and email addresses of all the people to whom you are likely to send email. Instead of having to type in the address when writing an email, you just click on the email address or addresses in the address book.

The screenshot shows an address book. Rather than type in the email address of the recipients and maybe make mistakes, you can simply click on their address. Notice the facility to create groups.

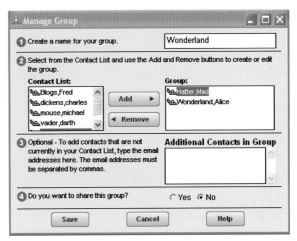

Using the advanced features of email, you can create groups.

Creating and sending an email to a group
In this activity you will learn the following skills:

» Manage contact details using an address book
» Create and use groups

For this activity you are required to produce a distribution list so that the same email can be sent to a group of people. You will need to find out how to do this yourself. Send the same email to four of your friends and check that they all received it.

Managing contact lists effectively

There are a number of email facilities that help save time when sending emails to more than one person and these are outlined below.

Groups

Groups are lists of people and their email addresses. They are used when an email needs to be distributed to people in a particular group. For example, if you were working as part of a team and needed to send each member the same email, then you would set up a group. Every time you needed to send the members of the group email, you could then just send the one email to the group thus saving time.

Using cc (carbon copy)

cc means carbon copy and it is used when you want to send an email to one person but you also want others to see the email you are sending. To do this you enter the email address of the main person you are sending it to and in the box marked cc you enter all the email addresses, separated by commas, of all the people you wish to receive a copy.

Using bcc (blind carbon copy)

bcc means blind carbon copy and this is useful when you want to send an email to one person and others but you do not want the others to see each other's email addresses. bcc is used by companies to hide email details when sending out multiple copies to prevent spam email and also to comply with the Data Protection Act.

Creating and managing an address book

An address book is part of an email package and it is here you can enter all your contacts' details such as names, addresses, telephone numbers, and email addresses. Setting this up takes a little time but once it is set up it is easy to send emails because you only have to click on the name, and the email address is automatically entered.

You can also set the address up so that everyone who sends you an email has their address automatically added to your address book.

Maintaining, storing and deleting emails

Rather than keep all the emails together it is better to be organized and set up folders. The way this is done depends on the email package you are using. When sending emails you can organize them in a similar way. For example, you could keep your personal emails in a separate folder to school/college-related work. It is important to remember to back up emails, as people tend to forget to do this.

Emails often need to be referred to in the future. To help with this task it is important to delete emails that are no longer needed. This will make finding remaining emails easier. It is also important in networks because you only have a limited amount of storage for your emails.

It is also necessary to delete any spam emails (i.e., unwanted advertising emails) which have been captured by the spam filter. In doing this you need to see if any important emails have become trapped in the spam filter.

Spam is a nuisance because you have to waste time checking to see if files you want are trapped in the filter and then delete the rest.

Sending other files electronically

There are a number of different ways of sending files from one place to another by using computers. The most popular way is by attaching a file to an email and then sending the two together.

File attachments

You can attach files to emails. For example, you could attach a file containing a photograph of yourself obtained from a digital camera, a piece of clip art, a picture that you have scanned in, a long document, etc. Basically, if you can store it as a file, then you can attach it to email.

You can attach more than one file to email, so if you had six photographs to send, then you could attach them and send them. Before you attach a file you must first prepare an email message to send, explaining the purpose of your email, and also giving some information about the files that you are sending (what their purpose is, what file format they are in, etc.). Once the email message has been completed, you click on the file attachment button and select the file you want to send. A box will appear to allow you to select the drive, folder, and eventually the file that you want to send.

If you want to send more than one file, you can select a group of files and attach them. Usually, if there are lots of files to send, the files will be compressed to reduce the time taken to send them.

Planning a structure for storing email messages
Emails are no different than other files. You need to create a folder structure to store them. This makes them easier to find and also to copy or delete them.

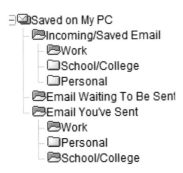

A folder structure has been set up here for incoming/saved emails and emails you've sent.

KEY WORDS

Email (sometimes called electronic mail) is a system for sending messages from one user to another using the Internet.
File attachment a file sent at the same time as an email. It is an easy way of sending files such as word-processed documents, photographs, spreadsheets, etc.

Activity 9.2

Creating a planned structure for emails
In this activity you will practise the following skills:

▸▸ Store email messages using a planned structure
▸▸ Store files using a planned structure

When you go onto the email package to view your email you will see that there is the option to create folders.

For this activity you have to plan a suitable folder structure to hold emails. It is important to note that you can create folders within folders (called subfolders).

Create this structure using the email software you use. Use the help facility provided by the software if you cannot work out how this is done. When you have created your structure you should move your existing emails into their appropriate folders.

Activity 9.3

Practising the use of email facilities
In this activity you will practise the following skills:

▸▸ Create new messages
▸▸ Use groups
▸▸ Use file attachments

Practise the use of some of the techniques covered in this chapter. Although all email software is slightly different, it all includes the facilities covered here.

You should check that you can create an address book and set up groups. Also try attaching files so that you can send them to yourself at your home email address. Attaching your files at school/college and sending them to your home email address is a good way of backing up your work.

Sending and receiving documents and other files electronically

Email software is very similar so do not worry if the email you use does not look exactly like this. All the principles of emails are the same. The email software being used here is a free email that you can set up from Google called Gmail.

You will have to take your teacher's advice on which email software you will be using for your course.

Activity 9.4

Setting up a Gmail account
In this activity you will learn the following skills:

▸▸ Locate specified information from a given website URL

If you want your screens to look exactly the same as those shown here you will need to set up a Gmail account. This is easy and free.

1 Log onto the Internet and enter the URL (i.e., web address) www.google.com.

The home page will appear similar to this:

Click on the Gmail link.

2 The opening Gmail screen is seen, where you can create a new email account.

Click on the "Create an account" button.

3 You now have to follow the instructions and enter the required information on the screen form that appears.

Creating email

In this activity you will learn the following skills:

▶▶ Send and receive documents and other files electronically

1 Access your email account. If you are using Gmail or other types of email you may be asked to supply some details first such as Username and Password. You may not be asked for these as you may have provided these when you first logged onto the Internet.

2 Once logged in the opening email screen appears.

To create an email click on Compose mail (this is called Write mail in some email packages).

3 You now see the new message window where you write your email.

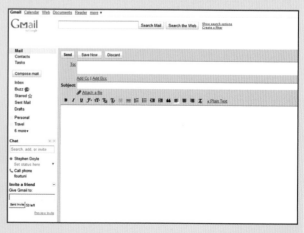

There are some important parts here:

The email address of the person you are sending the email to is entered here.

The subject line – contains a line on the subject of the email. Important for finding the email at a later date.

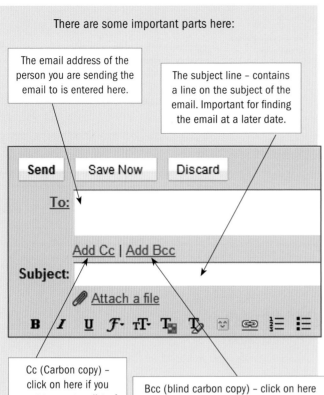

Cc (Carbon copy) – click on here if you want to create a list of email addresses who you want the same email to be sent to.

Bcc (blind carbon copy) – click on here to create a list of email addresses who you want the same email sent to. You do not want the other people to be able to see each other's email addresses.

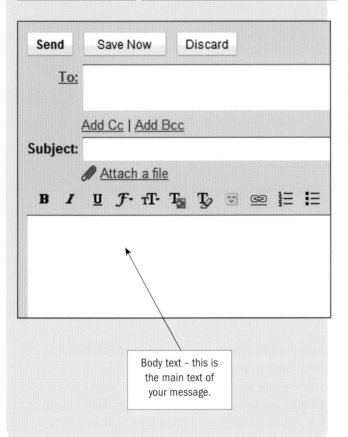

Body text – this is the main text of your message.

Making effective use of the Internet as a source of information

The Internet is the largest store of information in the world and being able to access relevant information quickly is an important skill which you will need to demonstrate in the examination.

There are a number of different ways of locating information using the Internet and these are described here.

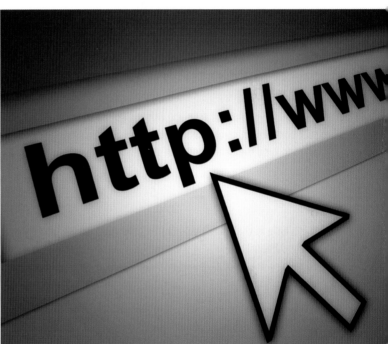

Many URLs start this way.

Locating specified information from a given web site URL

URL stands for Uniform Resource Locator, which is a complicated way of saying a web site address. One of the ways of accessing a web site is to type the URL (i.e., web site address) into a web browser or search engine. Each page of a web site has its own URL and the URL quoted in books/magazines, etc., will usually take you to the homepage of the web site. You can then search within the web site for the information you need by making use of links and search facilities.

Finding specific information using a search engine

Searching for information can be performed using a search engine such as Google or Yahoo. The search results are displayed on the basis of relevance or who has paid the most money to get in the top results. Clicking on a search result takes you to a web site and may take you straight to the information you want or you may go to the home page of the web site. It may then be necessary to perform a search of the content of the web site. This is done by entering key words into a search box.

A key word search on a website.

Complex searches

Don't be put off by the name. Complex searches help you save time searching for information, so they are worth knowing about. When you do a simple search, you may be overwhelmed by all the information. Complex searches help you narrow down a search.

AND

If you type in the search **USA AND flag** you will get all those documents that contain both words.

If you just type in **USA Flag** you will still get all those documents that contain both words. With most search engines you do not need to type the "and" between the words.

AND means: "I want **only** documents that contain **both** words."

OR

If you type in the search **USA OR flag** then you will get all the documents containing the word **USA**, all the documents containing the word **flag** and all those documents that contain both words.

OR means: "I want documents that contain either word. I don't care which word".

Searching for an exact match

If you want an exact match of words (i.e., the words side-by-side and in the same order), then put quotation marks around the words like this:

"Recipe for a chocolate chip cookie"

NOT

Suppose you want to search for information about different pets but you can't stand cats. You can exclude cats like this:

Pets NOT cats

Searching for a quotation

If you know the exact wording of the quotation, you can type it into the search engine. To get the exact match you need to put quotation marks around the words like this:

"I have a dream"

An advanced search page — notice that you can select the language for the web pages.

Activity 9.6

Fact detective

In this activity you will practise the following skills:

▸▸ Find specified information using a search engine

▸▸ Refine searches using more advanced search techniques

How good are you at tracking down information? Here are some things you are going to find out using the Internet and the search engine of your choice. Good luck!

1 The name of the animal with the Latin name Buffo Buffo.
2 The name of the plant that the drug called digitalis comes from.
3 Another name for the bone in the human body called the patella.
4 The name of the country that won the 2010 Football World Cup.
5 The names of the seven Ancient Wonders of the world
6 The date when man first walked on the Moon.
7 The name of the country that produces the most coffee in the world.
8 The four colours in the flag of Mauritius.
9 The name of the fastest animal in the world on land.
10 The name of the country that uses the currency called Ringgit.

Downloading and saving information obtained from web sites

You frequently have to download files from web sites and save them on your own computer. In many cases these files will be in a format called PDF as most computers come with a program capable of displaying these files.

Activity 9.7

Downloading files from a web site and saving them on a computer

In this activity you will learn the following skills:

▶▶ Download and save information as specified

▶▶ Store files using an appropriate planned structure

1   Access the Internet using your web browser and access the following web site by typing in the following URL:

http://www.cie.org.uk/qualifications/academic/middlesec/igcse/subject?assdef_id=969

This will access the home page of the University of Cambridge International Examinations IGCSE in ICT.

This page will look similar to this one:

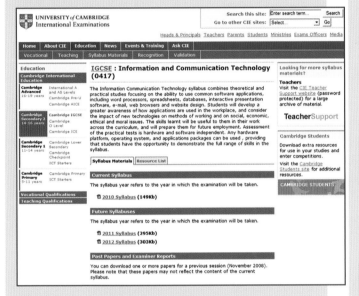

2   You will notice that there is a current syllabus and two future syllabuses. You will need to look for the syllabus for the year in which you will take the examination. Suppose you wanted the syllabus for 2012, you would click on the following:

📄 **2012 Syllabus (303Kb)**

You need to double left click on the correct syllabus for you.

Notice the little icon which tells you that the file is in PDF format.

The following window appears:

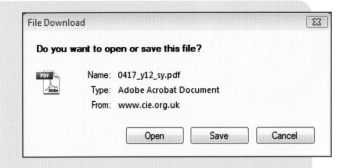

Notice that you can just view the file by clicking on open.

Click on Save.

3   You will now be able to choose where you want to save the file.

Locate your IGCSE ICT folder or where you store your IGCSE ICT files.

Notice that the filename is not very meaningful. It is important to use meaningful filenames as it makes them easier to find later.

File name: **0417_y12_sy**

Save as type: **Adobe Acrobat Document**

Delete the file name shown and replace with a filename similar to the following (NB you should change the year to the one you are using):

File name: **IGCSE Syllabus for 2012**

Save as type: **Adobe Acrobat Document**

Click on to save the file.

Save

4   Now locate the file and click on it to open it.

Activity 9.8

Downloading past papers

In this activity you will learn the following skills:

▶▶ Download and save information as specified

▶▶ Store files using an appropriate planned structure

For this activity you need to download CIE IGCSE past papers for you to refer to for your revision. As in the previous activity, you should change the names of the files so that they are more meaningful and easily found later.

10 Document production

Nearly everyone has a use for word-processing software for the production of documents such as letters, CVs, notes, simple leaflets, etc. You must build your skills in the use of word-processing software for the examination. This use goes beyond what you probably already know, so make sure you complete all the activities in this chapter.

The key concepts covered in this chapter are:
▶▶ Entering and editing data from different sources
▶▶ Organizing the page layout
▶▶ Formatting the text
▶▶ Ensuring the accuracy of the text

Word-processing packages

Word-processing software is ideal for the production and editing of text. You can type text into other software such as desktop publishing software, but it is easier to use word-processing software and then save it as a file which can be imported into other software.

Word-processing software has some the following advantages:

▶▶ Most people are very familiar with word-processing software so it saves time learning editing features of other software such as DTP software.
▶▶ You have the facility of mail merge which gives the ability to personalize documents such as letters, invitations, business cards, etc.
▶▶ It is very easy to import text from word-processing software into other software such as DTP and web authoring software.

MS Word is by far the most popular word-processing package, which according to Microsoft has over 500 million users. There are other word-processing packages available including:

▶▶ Corel WordPerfect
▶▶ OpenOffice (which includes word-processing software)
▶▶ Office Web Apps (which includes word-processing software)
▶▶ Google docs

The last two in the list are becoming very popular as they are web-based, which means you do not need to store the software on your computer as the software is accessed via the Internet.

Important note If you are asked a question, always remember to refer to "word-processing" software rather than give a brand name such as MS Word for which you will get no marks.

Enter and edit data from different sources

Create and open documents

Locating stored files

You will need to be able to locate stored files for the examination. You will also need to create folders for the various files you create. When saving files for the first time it is always best to use "Save As" because you can see where it is that the computer intends to save the file. If the location is not suitable then you can change it.

Before you start working through the activities, ensure you know where the files you need can be found. You will also need to decide where you will save any of the files amended or produced during the activities.

Using different file types

Not everyone uses the same piece of software and this means that data can be saved in different file formats. The good news is that despite them being in different file formats, we can usually open them using the software we are using to view and edit the information. Importing means bringing a file created using different software into the software you are currently using. For example, you might have a text file created using Windows Notepad and want to edit the file using word-processing software such as Microsoft Word.

Here are the file types you will need to know about for the examination.

CSV (comma separated value) file format

This is a list of data items that are separated from each other using commas. CSV files are mainly used for saving data from a table, spreadsheet or database where there are rows and columns of data. Comma separated value files use the file extension ".csv" after the filename.

Text files

As the name suggests, these files contain just text without any formatting. When these files are imported into the package you are using, you will have to edit the data in order to add the formatting. Text files use the file extension ".txt".

RTF (rich text format) files

This file format saves the text with a limited amount of formatting. Rich text format files use the file extension ".rtf".

Entering and editing text and numbers

Text and numbers can be entered into the word-processing software in a number of ways:

- using keyboard entry
- loading/importing a file created using the same word-processing software or created by another package
- copy and pasting from another document, web site, file, etc.

Using keyboard entry

If the text is not available to load/import or copy, then it is necessary to key it in using the keyboard. Keyboard entry needs to be done carefully in order not to introduce errors. Checking your work after typing is extremely important in the examination as the text and numbers you type in must exactly match the text you are instructed to type in.

Copy and paste

To copy text in a document, select it (the selected text will be highlighted) then right-click on the mouse. A menu will appear with a number of options including the following:

- Cut
- Copy
- Paste.

If you click on Copy, the text will now be placed in temporary storage (called the clipboard) by the computer. You can now move the cursor to another part of your document, where you want the text to be moved to. Right-clicking on the position brings up the menu again where you can choose to paste the contents into position. You will now have two copies of the text – one block in the old position and one block in the new position.

Cut and paste

To cut and paste, you simply select the text and right-click, and from the resulting menu select Cut. When the text is cut, the text is taken out of the document. You then click where you want the text to go, right-click and select Paste when the text appears in the new position.

Drag and drop

Drag and drop is a quick way of moving objects. You can drag and drop objects such as a section of text, file, photograph, piece of clip art, etc., onto a document. You can also drag and drop a file into a location (e.g. a folder or the Recycle bin).

Activity 10.1

Creating a table using data from a csv file

In this activity you will learn the following skills:

- How to import a file in csv file format into word-processing software
- How to place text into a table automatically
- How to make text bold

1 Load the word-processing software Microsoft Word and open the file **Countries**

(which has been saved in csv file format) by double clicking on the file icon.

Your teacher will tell you where to find this file.

When looking for this file in the folder, make sure that you select "All Files" otherwise only Microsoft Office files will be shown.

2 Notice the way the file is displayed. The first line contains column headings and other lines contain the actual data about the African countries. Notice also that all the items are separated by commas hence the name of the file format (comma separated values).

```
Name of country,Population,Capital
Ethiopia,85,Addis Ababa
Kenya,39,Nairobi
Algiers,33,Algiers
Ghana,23,Accra
Morocco,34,Rabat
Mozambique,20,Maputo
Nigeria,155,Abuja
Somalia,10,Mogadishu
South Africa,47,Pretoria
Sudan,39,Khartoum
Sengal,12,Dakar
Tanzania,38,Dodoma
Tunisia,10,Tunis
Uganda,28,Kampala
Zambia,15,Lusaka
Zimbabwe,13,Harare
```

3 Left click on the start of the data and keeping your finger pressed down on the left mouse button drag down until all the data is selected. When it is selected it will show blue like this:

```
Name of country,Population,Capital
Ethiopia,85,Addis Ababa
Kenya,39,Nairobi
Algiers,33,Algiers
Ghana,23,Accra
Morocco,34,Rabat
Mozambique,20,Maputo
Nigeria,155,Abuja
Somalia,10,Mogadishu
South Africa,47,Pretoria
Sudan,39,Khartoum
Sengal,12,Dakar
Tanzania,38,Dodoma
Tunisia,10,Tunis
Uganda,28,Kampala
Zambia,15,Lusaka
Zimbabwe,13,Harare
```

4 You are now going to show this data in a table. We do not need to work out the number of columns and rows in the table as the computer can do this automatically.

Click on ⌄ Tables in the toolbar. The following menu appears from which you should select the following:

- Insert Table...
- Draw Table
- Convert Text to Table...
- Excel Spreadsheet
- Quick Tables ▶

5 The following box appears:

Notice that the number of columns and rows has been determined by the data itself.

Click on "OK" to put the data into the table. Your table should now appear like this:

Name of country	Population	Capital
Ethiopia	85	Addis Ababa
Kenya	39	Nairobi
Algiers	33	Algiers
Ghana	23	Accra
Morocco	34	Rabat
Mozambique	20	Maputo
Nigeria	155	Abuja
Somalia	10	Mogadishu
South Africa	47	Pretoria
Sudan	39	Khartoum
Senegal	12	Dakar
Tanzania	38	Dodoma
Tunisia	10	Tunis
Uganda	28	Kampala
Zambia	15	Lusaka
Zimbabwe	13	Harare

Select the first row containing the column headings and then click on **B**.

This makes the column headings bold.

6 Save this word-processed file using the filename **Table of countries**

Activity 10.2

Entering and editing text

In this activity you will learn the following skills:

- ▶▶ Accurately use the keyboard to enter text
- ▶▶ Use drag and drop to move text around
- ▶▶ Alter the font size
- ▶▶ Centre text
- ▶▶ Change the line spacing in a document
- ▶▶ Copy and paste an object (i.e. a table in this case) from one document to another
- ▶▶ Alter the border on a table
- ▶▶ Import an image file and position it in a document
- ▶▶ Edit an image by cropping it

1 Load the word-processing software Microsoft Word and create a new document. Key in the following text:

Africa is the world's second largest continent according to land size as well as by size of population. The African continent consists of 54 separate countries and about 14% of the world's total population live here.

The climate in Africa varies from tropical to subarctic on its highest peaks. In the northern part of Africa there are deserts and most of the land is arid. In the central and southern parts there are savannah plains and dense rainforests.

Africa boasts the greatest variety of animals and includes animals such as lions, elephants, cheetahs, deer, giraffes, camels, monkeys, etc. One ecological problem is deforestation where forests are being destroyed to make room for the growing of crops. This is destroying the habitat for many of the endangered species.

Most scientists regard the African continent to be the origin of the human species and evidence has been found of the first modern human called Homo sapiens dating to around 200,000 years old.

2   Carefully proof-read the text you have typed in by comparing it with the original text. The two should be identical. If there are any errors, these should be corrected before proceeding.

3   Select the entire third paragraph by highlighting it. Once selected, click on the highlighted text and keeping the right mouse button pressed down drag the cursor to the line between the first and second paragraphs. This technique is used to move text around.

Your text should now look like this:

Africa is the world's second largest continent according to land size as well as by size of population. The African continent consists of 54 separate countries and about 14% of the world's total population live here.

Africa boasts the greatest variety of animals and includes animals such as lions, elephants, cheetahs, deer, giraffes, camels, monkeys etc. One ecological problem is deforestation where forests are being destroyed to make room for the growing of crops. This is destroying the habitat for many of the endangered species.

The climate in Africa varies from tropical to subarctic on its highest peaks. In the northern part of Africa there are deserts and most of the land is arid. In the central and southern parts there are savannah plains and dense rainforests.

Most scientists regard the African continent to be the origin of the human species and evidence has been found of the first modern human called Homo sapiens dating to around 200,000 years old.

4   In the next to last paragraph delete the sentence: "This is destroying the habitat for many of the endangered species." This should be done by highlighting the sentence and then clicking on ✂ Cut . This is the quickest way of deleting a section of text.

Check that you have a line between each paragraph, otherwise insert one.

5   The document needs a title. Insert the title "Africa" at the top of the page. You will need to move the existing text down two lines to leave room for this title. Do this by clicking on the first letter of the first sentence and press enter twice.

Select the title by highlighting it.

Africa

Africa is the world's second largest continent according to land size as well as by s of population. The African continent consists of 54 separate countries and about 1 of the world's total population live here.

Click on the font size and change it to 20pt like this

20 ▾ .

Make the text bold by clicking on **B** and centre the text by clicking on ≡

6   You are now going to increase the line spacing between the lines of text. Select all the text other than the title by highlighting it.

Ensure you have the Home tab selected and then from the paragraph section click on ↕≡▾ .

From the resulting list, select the line spacing 1.5 like this:

✓	1.0
	1.15
	1.5
	2.0
	2.5
	3.0
	Line Spacing Options...
≡	Add Space Before Paragraph
≡	Add Space After Paragraph

The document with the increased line spacing is now shown.

7   You are now going to add the table created in the previous activity.

Click midway between the first and the second paragraph as this is where the table is to be inserted.

Open the file containing the table called **Table of countries**

Select the entire table by highlighting it like this:

Name of country	Population	Capital
Ethiopia	85	Addis Ababa
Kenya	39	Nairobi
Algiers	33	Algiers
Ghana	23	Accra
Morocco	34	Rabat
Mozambique	20	Maputo
Nigeria	155	Abuja
Somalia	10	Mogadishu
South Africa	47	Pretoria
Sudan	39	Khartoum
Senegal	12	Dakar
Tanzania	38	Dodoma
Tunisia	10	Tunis
Uganda	28	Kampala
Zambia	15	Lusaka
Zimbabwe	13	Harare

Click on 📋 Copy which copies the selected area to a storage place called the clipboard.

Go back to the document about Africa and click on

Paste which puts the table into the correct position.

8 The table needs adjusting to line it up with the text.

Left click on the left table border like this and keeping the mouse button pressed down drag the left border to the right so that it is lined up with the text. Now line up the right margin with the text.

Your document will now look like this:

Africa

Africa is the world's second largest continent according to land size as well as by size of population. The African continent consists of 54 separate countries and about 14% of the world's total population live here.

Name of country	Population	Capital
Ethiopia	85	Addis Ababa
Kenya	39	Nairobi
Algiers	33	Algiers
Ghana	23	Accra
Morocco	34	Rabat
Mozambique	20	Maputo
Nigeria	155	Abuja
Somalia	10	Mogadishu
South Africa	47	Pretoria
Sudan	39	Khartoum
Senegal	12	Dakar
Tanzania	38	Dodoma
Tunisia	10	Tunis
Uganda	28	Kampala
Zambia	15	Lusaka
Zimbabwe	13	Harare

Most scientists regard the African continent to be the origin of the human species and

---

9 You are now going to insert a digital image into the document.

Click midway between the fourth and fifth paragraph.

Click on the [Insert] tab and then select Picture.

You now need to locate the file called Map of Africa in the area where all the files you need are stored. Double left click on the file to insert the image into the document.

10 This image shows Europe as well as Africa and it would be best to only use part of the image (i.e. mainly Africa). You are going to cut out part of the image and discard the remainder. This is called cropping the image.

Select the image by clicking on it. You will see the picture formatting toolbar at the top of the screen.

Notice the size section where you can alter the height and width of an image as well as crop an image.

Click on the Crop icon.

The image will appear like this with the black handles that can be used to crop the image around the edges of the image.

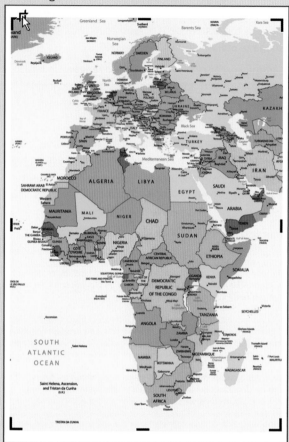

Use the handles to drag the corners and edges until the image looks a bit like this:

11 The article is now complete so save it using the filename **Completed Africa article**.

Activity 10.3

Placing and manipulating images

In this activity you will learn the following skills:

▸▸ Locating stored files

▸▸ Importing an image file

▸▸ Sizing an image maintaining the aspect ratio

▸▸ Wrap text around an image

▸▸ Adjust a margin

1   Load the word-processing software and open the document called **Distributed computing**

Check you have the right document displayed by comparing the start of the document with that shown here:

Distributed computing using the Internet

One main problem in collecting data is that there is so much of it to analyse. Analys
amounts of data requires a lot of computer power and this power may not be availab
because of the cost.

In many cases this problem can be solved by using distributed computing using the
Here, instead of using a huge expensive supercomputer to do the job, it could be do

---

2   There are three images to be put into this text and they have been saved using the following filenames:

Dist_computing_image1
Dist_computing_image2
Dist_computing_image3

All these images have been saved in the folder for this book.

Position the cursor on the line between the first and second paragraphs of the document.

Click on the insert tab | Insert | and then click on | Picture |.

You will then have to locate the folder containing the image files that are to be imported into this document.

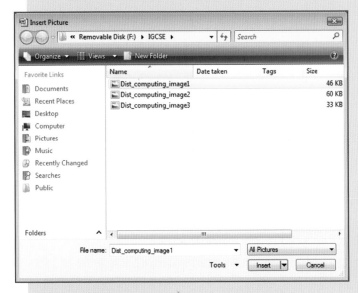

Once found, click on the file Dist_computing_image1 and then click on the Insert button.

The image is inserted into the document like this:

3   Right click on the image and from the menu that appears choose Size when this dialogue box appears:

Notice the tick in the Lock aspect ratio box. This means that if either the height or width of the image is changed then the other dimension will change in order to keep the image in proportion. If you want to change the size of the image in one direction only, then this tick will need to be removed.

At the moment the image is too large, so reduce the height of the image from 21.56 cm to 13cm.

Click on Close.

4 Right click on the image again and this time select Text Wrapping from the menu.

You will then be presented with a number of text wrapping options:

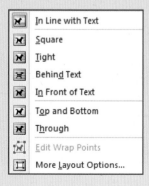

Notice these text wrapping options. You might be asked to use any of the following:

- ▶▶ Square – if the image is placed in the centre of text the text flows around the image leaving a square shaped image.
- ▶▶ Tight – here the text moves in to the shape of the image on the right- and left-hand side of the image. You can only see this effect if the image is not a regular shape.
- ▶▶ Top and bottom (this is called above and below in the CIE specification) and this places the image with the text above and below the image but with no text wrapped to the side of the image.

Click on the option "Square" in the menu and the text will appear like this:

Distributed computing using the Internet

One main problem in collecting data is that there is so much of it to analyse. Analysing large amounts of data requires a lot of computer power and this power may not be available because of the cost.

In many cases this problem can be solved by using distributed computing using the Internet. Here, instead of using a huge expensive supercomputer to do the job, it could be done using a few less powerful computers connected together using the Internet but with each working on the same problem. In many cases computers just sit around not doing much for most of the time so why not use their processing power?

Some distributed systems ask home users to contribute some of their computing resources. For example, there is climate change research going on at the moment that requires a huge amount of data processing. Home users can contribute some of their wasted computer time to this project.

Advantages of distributed computing

- reduces cost because an expensive powerful computer such as a supercomputer is not needed

5 It is necessary to tidy the text up. Click just before the "Advantages of distributed computing" and press the enter key twice to move this heading down.

6 Directly under the image, type in the following text using the font Times New Roman and a font size of 10pts:

Distributed computing is a very useful tool for scientists who have huge amounts of data to analyse when trying to model systems such as the Earth's climate

This text is a caption for the image, so it would look better if it were lined up with the edges of the image. You need to adjust the right margin to do this.

Highlight the section of text you have just typed in.

Left click on the triangle for the right indent `14 · 1 · 15 · 1 · △` and holding the mouse button down, drag the margin to the left until it is lined up with the right-hand edge of the image.

The caption now looks like this:

Distributed computing is a very useful tool for scientists who have huge amounts of data to analyse when trying to model systems such as the Earth's climate

It would improve the appearance if the text was fully justified. This means that the text is lined up to both the right and left margins.

With the text still highlighted click on the ▤ (Justified) button.

The text is now fully justified like this:

Distributed computing is a very useful tool for scientists who have huge amounts of data to analyse when trying to model systems such as the Earth's climate

7 You are now going to insert another image from file.

Click between these two lines shown by the vertical line here, as this will be the position where the image is to be inserted.

this would provide evidence of extraterrestrial life.

The problem the project has is that there is a lot of background noise which includes radio

8 Click on the insert tab `Insert` and then on `Picture`.

You will then have to locate the folder containing the image file

Dist_computing_image2

Once found, click on the file Dist_computing_image2 and then click on the Insert button.

The image is inserted into the document like this:

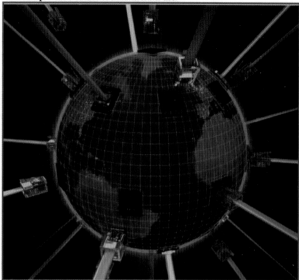

signals from outer space. If the signals were confined to a narrow range of frequencies, then this would provide evidence of extraterrestrial life.

The problem the project has is that there is a lot of background noise which includes radio signals from TV stations, radar, satellites and from celestial sources. It is very difficult to analyse the data from radio telescopes and look for other signals that could indicate extraterrestrial life.

9 Re-size the document so that the image is 60% of its original size and in the same proportions (i.e. make sure that the "Lock aspect ratio" box is ticked).

Now wrap the text so that it is square with the image.

When these actions have both been completed correctly the section containing the image will look like this:

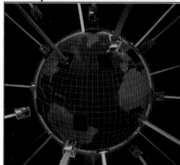

signals from outer space. If the signals were confined to a narrow range of frequencies, then this would provide evidence of extraterrestrial life.

The problem the project has is that there is a lot of background noise which includes radio signals from TV stations, radar, satellites and from celestial sources. It is very difficult to analyse the data from radio telescopes and look for other signals that could indicate extraterrestrial life.

In order to search for the narrow-bandwidth signals lots of computing power is needed. At first supercomputers containing parallel processors were used to process the huge amount of the data from the telescopes. Then someone came up with the idea of using a virtual supercomputer consisting of a huge number of Internet-connected home computers. The project was then named SETI@home and it has been running since 1999.

At the time of writing the SETI@home project uses 170,000 active volunteers around the world and uses 320,000 computers but even with this power they still need more!

10 Insert the image Dist_computing_image3 in the position shown.

Click on the image and use the handles on the image to re-size it so that the right edge of the image is lined up with the right margin for the text.

Your image should be in the position as shown here:

home computers. The project was then named SETI@home and it has been running since 1999.

At the time of writing the SETI@home project uses 170,000 active volunteers around the world and uses 320,000 computers but even with this power they still need more!

11 Save your document using the file name **Dist_computing_final**

Including different formats of information from the Internet
You often want to copy information from the Internet and use this information in your own documents. Such information may include:

▸▸ text
▸▸ images
▸▸ tables
▸▸ graphs and charts.

In most cases the information you need will be embedded in a web page. In these cases all you need to do is highlight the object in the case of text and tables or simply click on the object in the case of images and graphs/charts. You can then right click and select copy, which copies your selection into the clipboard. You then position the cursor on your document to mark the insertion point and right click the mouse again and select Paste. The object will then appear in your document where you can then edit it.

Organizing the page layout

The page layout is the arrangement of the items (text, images, etc.) on the page of the document. Here are some of the things that are regarded as part of the page layout. Most of these can be altered by pressing the Page Layout tab: **Page Layout**

Once the Page Layout tab has been pressed, you are presented with many tools that help you lay out the page and the main ones are shown here:

Size Setting the page size (this is usually determined by the size of the paper you are printing the document on). Page sizes include A4, A5 (which is half the size of A4), letter, etc.).

Orientation Page orientation can be portrait or landscape

Portrait

Landscape

Margins Here you can adjust the size of the margins around the page (i.e. right, left, top and bottom). The margins define where the main text and other objects such as images can be placed. You can also set gutters which allow extra space to allow the pages to be bound without concealing any information on the page.

Columns Using the column tool you can put the text into columns. You can also set up the width of each column and the space between the columns and even have a vertical line or lines separating the column.

Breaks ▾ Using breaks you can force the computer to start a new page rather than allow it to only start a page when the page is completely filled. You can insert section breaks and column breaks so that the text can run on from one section or column to the next. Breaks also allow you to prevent the occurrence of widows and orphans which you will learn about later.

Some page layout tools are accessed from the Insert tab **Insert** and these include:

Header Footer Here you can set headers (at the top of the page) or footers (at the bottom of the page). These can contain items such as date, page numbers, name of document author, name of the document file, etc.

There are number of other page layout tools that are accessed by first clicking on the **Home** tab.

Clicking on this button brings down a menu allowing you to adjust the line spacing between lines of text in a document. There are various options such as 1.0, 1.15, 1.5, 2.0, 2.5, 3.0. You can also use this to add lines automatically before or after a paragraph.

These buttons are used for aligning sections of text. Each button is used to align text in the following way:

Aligns text to the left.

Centres the text.

Aligns text to the right.

Aligns the text to both the left and right margins – called fully justified text.

Used to decrease or increase the indent level of a paragraph.

The following appears no matter which tab has been selected:

This is the ruler which contains a number of tools. On the left hand side of the ruler there are the following three sliders that can be moved along the ruler to control the way text appears on the page:

This is used to indent the first line by a certain amount.

This is used to left indent the text.

This is used to provide a hanging indent.

Look carefully at the following, which shows how the text appears when these indents are used.

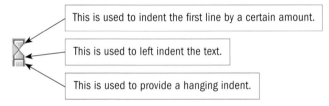

Feasibility report
Feasibility is an initial in

In the above, a paragraph indent of 0.5 cm has been applied to the first line of a paragraph.

Feasibility is an importa
not developed that have

In the above the whole section of selected text has been indented by 1 cm.

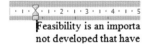

Feasibility is an import;
are not developed that have

In the above the main body of text has been indented by 1.5 cm

from the left margin and the first line of the paragraph has been indented by 0.75cm from the line of the main text.

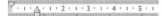

Analysis
Analysis looks in detail at the cur
never been performed befor
systems analyst or just analy

This setting produces a hanging indent of 1cm. The first line is lined up with the margin and the other lines are lined up using the hanging indent.

Ensuring that the page layout is consistent

It is important to create professional-looking documents. It is therefore essential that the page layout is consistent on each page and for all the pages in a multi-page document. Page layout consistency needs to be considered when creating the document and also during the proof-reading stage.

In the examination you will need to ensure that the page layout is consistent and you should ensure that each of the following has been used consistently:

▶▶ Font styles – ensure that headings use the same font style, subheadings have the same font style and the main body text has the same font style unless you are instructed otherwise.
▶▶ Alignment – text can be left aligned, centred or right aligned. It can also be justified, which means the text is aligned to both the left and the right margins. Ensure that you have obeyed any specific instructions in the examination and have maintained consistency.
▶▶ Spacing between lines, paragraphs and before and after headings – if you decide to leave two blank lines after a main heading, then there should be consistency whenever there are other main headings. All other spacing needs to be consistent.

Activity 10.4

Setting the page layout for a document
In this activity you will learn the following skills:

▶▶ Import a document which is in rich text format (.rtf) into your word processor
▶▶ Set the margins for a document
▶▶ Set the paper size for a document
▶▶ Insert headers and footers in a document
▶▶ Add page numbers, dates and filenames automatically
▶▶ Set the alignment of text in headers and footers

For the examination you will need to set the page layout for a document. You will need to follow the instructions given carefully.

1 Load the word-processing software Microsoft Word and open the document called **Whizz Logistic**

➡

This file has been saved in rich text format and this file can be loaded into Word and needs no alteration as both the text and the simple formatting in this document are kept.

2 Click on the | Page Layout | tab on the toolbar and select

Margins and from the items in the menu select "Custom Margins". The following box appears:

Notice the margin section where you can change the size of the margins of the page you are producing.

Set the top, bottom, left and right margins to 3 centimetres each. When you have done this correctly the margin section will look the same as this:

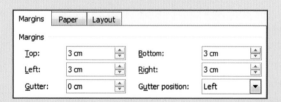

Ensure that in the orientation section "Portrait" has been selected.

3 Click on the Paper tab | Paper |. It is from here you can select the size of the paper to be used. Notice the paper sizes include A4, A5, Legal, etc. Check that A4 has been selected for this document.

Click on OK.

4 You are now going to insert headers and footers in the document. A header is an area on the page between the top margin and the top of the page. A footer is an area between the bottom margin and the bottom of the page. Headers and footers are used for such things as the name of the author of the document, the date the document was produced, page numbers, etc.

Click on the insert tab | Insert | in the toolbar and then

click on Header and then choose the following layout from the list.

In the examination you will need to follow instructions as to what to put into the header.

Place the following information in the header:

Your own name, which is to be left aligned, and the number 121121, which is to be right aligned.

Click on the left placeholder box when it will be highlighted like this:

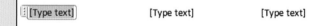

Type your name here. The text you type will replace the text already there.

Click on the right placeholder box and type in the number 121121.

Click on the centre placeholder box and click the backspace key to delete the box.

Your header should now look like this:

5 Click on the insert tab | Insert | in the toolbar and then

click on Footer and choose the following from the list:

You are going to include the following in the footer:

▸▸ An automated filename left aligned.

▸▸ An automated page number centre aligned.

▸▸ Today's date right aligned.

To create the automated filename click on the left placeholder box on the left-hand side of the footer and

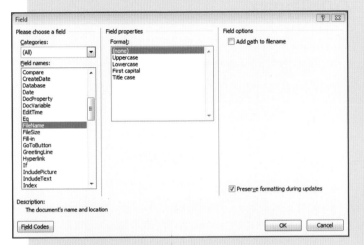

click on Quick Parts ▾ and then select Field from the menu. The following box appears from which you need to select FileName from the list of fields.

Click on OK to continue.

The filename now appears in the footer like this:

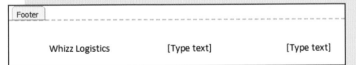

6 To insert the page number, click on the centre place

holder box and then click on Page Number ▾ and pick "Current Position" from the menu options. You are now presented with some options and you need to choose the one shown here:

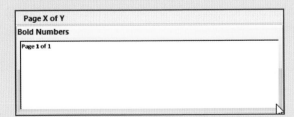

The page number now appears in the footer like this:

If the place holder box has forced onto the line below delete the place holder and then position the cursor after the text "Page 1 of 2".

Click on Date & Time and then select the date format highlighted as shown:

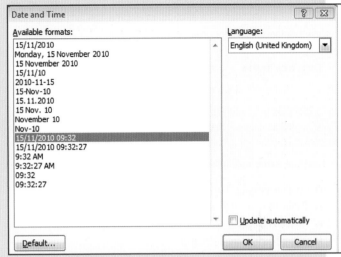

Click on OK.

Important note

Different countries write dates in different formats. When asked in the examination to insert a date, it does not matter which format you use.

The date and time is inserted next to the page number like this:

You now need to click just after the 2 of the page number and press the tab key until the end of the date-time is almost aligned with the right margin. To finish off you will need to press the space bar until the date-time is exactly aligned with the right margin.

The final footer should look like this:

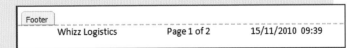

7 Save the file this time ensuring that it is saved still in rich text format.

8 Print a copy of the document.

Formatting the text

There are a number of things you can do to text to make it stand out more. In this section you will learn about formatting text.

Font type and font size

Changing the font type (e.g., Arial, Times New Roman, etc.) alters the appearance of the characters. Font types are given names and you can change the font by selecting the text and then clicking on the correct part of the formatting toolbar shown below. Notice also that there is a section for altering the font size (i.e., how big the characters appear).

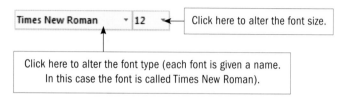

Click here to alter the font size.

Click here to alter the font type (each font is given a name. In this case the font is called Times New Roman).

Here are some examples of different font types:

- ▸ This text is in Times New Roman
- ▸ This text is in Arial
- ▸ This text is in Century Schoolbook
- ▸ This text is in Tahoma

Font size is measured in points or pt for short – 20pt text is much larger than 12pt text. You will often be given instructions as to what point size to use for headings, sections of text, etc., in the documents that you produce during the examination.

There is another way to alter the size of selected text. You can press either of these buttons to increase or decrease the font size.

Serif and sans serif fonts

There are two main types of font: serif fonts and sans serif fonts. It is very important for the examination that you know the difference between them.

Serif fonts are those fonts that have detail attached to the strokes that make up the letters. Sans serif fonts do not have this extra detail added. "Sans" in French means without, so sans serif means without serifs. Serif fonts were developed so that a reader's eye is led to the next letter in a large block of text.

A

This letter is in a serif font – notice the detail at the bottom of the strokes.

A

This letter is in a sans serif font – notice that there is no detail on the strokes. Sans serif fonts are usually better for titles, headings and posters.

Serif fonts	Sans serif fonts
Times New Roman	Arial
Garamond	Verdana
Century Schoolbook	Tahoma

Important note When preparing documents in the exam you will be asked to use a serif or sans serif font, so it is important to be able to spot one. Stick with the common fonts in the table.

Activity 10.5

Serif or sans serif

In this activity you will practise the following skills:
- ▸▸ Distinguish between serif and sans serif fonts

Here are some fonts. You decide whether each of them is a serif or sans serif font.

1   A
2   V
3   H
4   **Y**
5   K
6   F
7   U
8   X
9   R
10 R

Using an appropriate font

Most of the examination questions ask you to use specific font types (serif or sans serif) and they do not usually specify the font name. You therefore have to choose from a number of different fonts. You must make sure that whichever font you choose is easy to read.

Font styles

Font styles include bold (**text in heavy type**), italic (*text slanting to the right*) and underlining (<u>text with a line underneath</u>). These can be used to highlight certain words or text in order to draw particular attention to them.

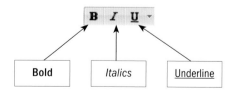

Bold   *Italics*   <u>Underline</u>

Highlighting text

Just like you can highlight printed text using a highlighting pen (usually in a bright colour), you can produce highlighted text in a word-processed document.

To highlight text you first need to select it like this

This text is going to be highlighted.

Then click on 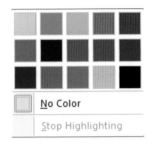.

You will then be presented with a palette of colours to choose from.

When you click on one of these colours the text is highlighted like this:

This text is going to be highlighted.

Formatting lists

Bullets

Bullets are useful to draw the reader's attention to a list of points you would like to make. Rather than just write an ordinary list, you can put a bullet point (a dot, arrow, diamond, square or even a small picture) at the start of the point. The text for your points is then indented a small amount from the bullet.

To produce a bulleted list:

1   Select the section of text you want bulleted by highlighting it.
2   Click on the button ☰ ▾.
3   The list will be bulleted using the default bullet (i.e. black dots) like this:
   • Point 1
   • Point 2
   • Point 3

4   You can choose a different bullet. After highlighting the text you can click on the drop down arrow on the bullet button. This menu appears from which you can choose a different type of bullet.

Numbered lists

Numbered lists are useful if the listed items need to be put into a particular order.

1   Select the section of text you want put into a numbered list by highlighting it.

2   Click on the button ☰ ▾.

3   The list will be bulleted using the default numbers like this:

   1   First point
   2   Second point
   3   Third point

4   You can choose a different number style. After highlighting the text you can click on the drop down arrow on the numbered list button. This menu appears, from which you can choose a different type of number such as Roman numerals. Notice also that you can have a lettered list.

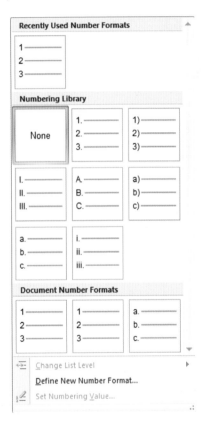

Creating and editing tables

Tables are made up of rows and columns that can be filled with text. Tables are used to organize and present information. They may also be used to align numbers in columns and once this is done you can perform calculations on them. There are many different ways of creating a table and one way in Word is to use the Table menu. To make creating tables easy there are set formats already created and you can just select the table format that suits your needs.

Merge and split table cells – if you create a table and then want to type a heading in the first cell, this is what happens:

This is the heading and you can see what happens if you type in too much				

The text is wrapped in order to keep the width of the cell the same. This is not what you want to happen with a heading, so it is necessary to merge cells. When this is done the heading fits in like this:

This is the heading when the cells have been merged to fit the heading				

It is also possible to split cells like this				

Activity 10.6

Creating a table

In this activity you will learn the following skills:

▸▸ Create a table with a specified number of columns and rows

1 Click on the Insert tab | Insert |.

2 Click on | Table | Tables | and the following menu appears:

```
Insert Table
[grid of empty cells]

⊞  Insert Table...
✎  Draw Table
⊞  Convert Text to Table...
⊞  Excel Spreadsheet
⊞  Quick Tables              ▶
```

3 Tables consist of columns and rows. Suppose you wanted a table containing 5 columns and 6 rows. All you do is run the cursor so that you have 5 columns and 6 rows highlighted like this:

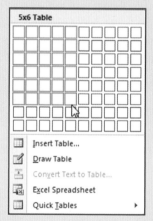

```
5x6 Table
[grid of cells with 5x6 highlighted]

⊞  Insert Table...
✎  Draw Table
⊞  Convert Text to Table...
⊞  Excel Spreadsheet
⊞  Quick Tables              ▶
```

When you have the correct number highlighted, left click and the blank table is inserted into the document like this:

4 Close the file without saving.

Activity 10.7

Editing tables

In this activity you will learn the following skills:

▸▸ Insert a table into a document

▸▸ Adjust the column width and row height

▸▸ Input data into a table

You are going to produce a table to show two of the main differences between a wide area network (WAN) and a local area network (LAN).

1 Load Microsoft Word and start a new document.

2 Set up a table consisting of 3 columns and 3 rows. Look back at the section on "Creating a table" if you have forgotten how to do this.

The following table will appear at the cursor position. Notice that it spans from one margin to the other.

Adjust the width of the columns and the height of the rows in the following way. Move the pointer onto the

lines and you will notice that the cursor changes to two parallel lines (either vertical or horizontal). Press the left mouse button and drag until the widths are similar to those shown in the following diagram.

	LAN	WAN
Difference 1	Confined to a small area, usually a single site.	Covers a wide geographical area spanning towns, countries or even continents.
Difference 2	Usually uses simple cable links owned by the company.	Uses expensive telecommunication links not owned by the company.

3 Type the data shown in the table above. You can centre the headings "LAN" and "WAN" by typing them, highlighting them and then clicking on the centre button on the toolbar. This will centre the headings in the column. Also, embolden the words "LAN", "WAN", "Difference 1" and "Difference 2".

4 See how neat this now looks. Save your document using the filename **The differences between a LAN and a WAN**. Print out a copy of your document.

Activity 10.8

Editing a table

In this activity you will learn the following skills:

▸▸ Insert rows, delete rows, insert columns and delete columns

▸▸ Set horizontal cell alignment

▸▸ Set vertical cell alignment

1 Load the word-processing software Microsoft Word and open the document called **Tours**.

Check that you have the following table loaded:

Tour	Duration (Hrs)	Adult cost($)	Child cost($)	Date
El-Alamein	6.5	45	27	12/12/11
Alexandria City	4	40	24	13/12/11
Cairo, Pyramids and Tombs	13	80	55	12/12/11
Classic Cairo	12.5	80	55	13/12/11
Caravans in the sand	13.5	110	70	12/12/11
The Pyramids and the River Nile	12	108	68	12/12/11
Cairo overland	48	250	145	12/12/11
Luxor & Valley of the Kings	14	300	175	12/12/11

2 Two extra tours have been added to the list:

▸▸ Landmarks of Alexandria for a duration of 5 hrs at a cost of $50 for adults and $35 for children on 13/12/11

▸▸ Alexandria panorama for a duration of 7 hrs at a costs of $100 for adults and $70 for children on 13/12/11

You need to add two blank rows to hold this data at the bottom of the list.

To do this click on the very last cell in the table.

Press the tab key once. You will see a blank row has been inserted.

Now click on the very last cell of the row you have just inserted and press the tab key to insert another blank row.

Now add the data shown above into these two rows.

3 You are now going to insert a column between the columns for the Child cost($) and the Date.

Click on the cell with the column heading "Child cost($)" in it and then right click the mouse button and the following menu appears:

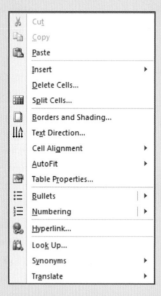

Look carefully at this menu and notice that it provides the tools for editing tables.

Click on "Insert" – the following choices appear:

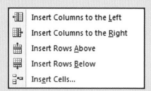

Click on "Insert Columns to the Right".

The blank column now appears in the correct position.

4 Key in the data as shown here. Make sure that the column heading is made bold.

Tour	Duration (Hrs)	Adult cost($)	Child cost($)	Meal provided?	Date
El-Alamein	6.5	45	27	N	12/12/11
Alexandria City	4	40	24	N	13/12/11
Cairo, Pyramids and Tombs	13	80	55	Y	12/12/11
Classic Cairo	12.5	80	55	Y	13/12/11
Caravans in the sand	13.5	110	70	Y	12/12/11
The Pyramids and the River Nile	12	108	68	Y	12/12/11
Cairo overland	48	250	145	Y	12/12/11
Luxor & Valley of the Kings	14	300	175	Y	12/12/11
Landmarks of Alexandria	5	50	35	N	13/12/11
Alexandria panorama	7	100	70	N	13/12/11

5 Insert a row between the tours "Classic Cairo" and "Caravans in the sand" by clicking on the row for "Classic Cairo" and then right clicking to show the menu and then click on Insert to insert the blank row below. Then enter the data as show here:

Medieval Cairo	13	95	60	Y	12/12/11

6 It has been decided to run each tour on both dates so the date column needs deleting. Right click on the "Date" and from the menu that appears choose delete cells when the following choices appear.

Choose "Delete entire column" and then click on OK to confirm the deletion.

7 The table will be improved if the contents of the cells are aligned. Highlight all the cells except the tour names like this:

Tour	Duration (Hrs)	Adult cost($)	Child cost($)	Meal provided?
El-Alamein	6.5	45	27	N
Alexandria City	4	40	24	N
Cairo, Pyramids and Tombs	13	80	55	Y
Classic Cairo	12.5	80	55	Y
Medieval Cairo	13	95	60	Y
Caravans in the sand	13.5	110	70	Y
The Pyramids and the River Nile	12	108	68	Y
Cairo overland	48	250	145	Y
Luxor & Valley of the Kings	14	300	175	Y
Landmarks of Alexandria	5	50	35	N
Alexandria panorama	7	100	70	N

Right click on the highlighted part of the table and select "Cell Alignment" and the following options appear:

This is a good place to place the cursor over each of the alternatives so that the caption as to what the button does appears. Here is a table show how the text will appear when each button is applied to cells containing text:

Align Top Left	Align Top Centre	Align Top Right
Align Centre Left	Align Centre	Align Centre Right
Align Bottom Left	Align Bottom Centre	Align Bottom Right

Select "Align Centre" – this will position the data centrally (horizontally and vertically) in the cells. Your table will now look like this:

Tour	Duration (Hrs)	Adult cost($)	Child cost($)	Meal provided?
El-Alamein	6.5	45	27	N
Alexandria City	4	40	24	N
Cairo, Pyramids and Tombs	13	80	55	Y
Classic Cairo	12.5	80	55	Y
Medieval Cairo	13	95	60	Y
Caravans in the sand	13.5	110	70	Y
The Pyramids and the River Nile	12	108	68	Y
Cairo overland	48	250	145	Y
Luxor & Valley of the Kings	14	300	175	Y
Landmarks of Alexandria	5	50	35	N
Alexandria panorama	7	100	70	N

8 You are now going to shade the cells containing the column headings to make them stand out more. Select them by highlighting them like this:

Tour	Duration (Hrs)	Adult cost($)	Child cost($)	Meal provided?

Write click on the shaded part and the following menu appears:

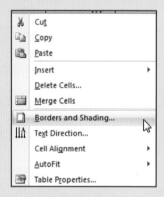

Click on "Borders and Shading" and the following dialogue box appears:

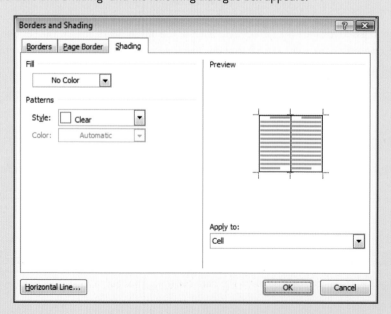

This box allows you to fill selected cells with a pattern. You can also add colour as the background to selected cells.

Click on "Style" and change it to the following:

Style: ▢ 15% ▾

This will shade the selected cells so the column headings look like this:

Tour	Duration (Hrs)	Adult cost($)	Child cost($)	Meal provided?

9 Save your new table using the filename **Final_tours_table**

Activity 10.9

Formatting a table

In this activity you will learn the following skills:

▶▶ Colour cells
▶▶ Hide the gridlines of a table

1 Load the software Microsoft Word and open the file **Stopping_distances**

This document shows a table showing the stopping distances for a car at certain speeds.

Speed (km/h)	Thinking distance (m)	Braking distance (m)	Overall stopping distance (m)
32	6	6	12
48	9	14	23
64	12	24	36
80	15	38	53
96	18	55	73
112	21	75	96

Select the column headings by highlighting them.

2 Right click on the highlighted area and select "Borders and Shading".

Click on the Shading tab:

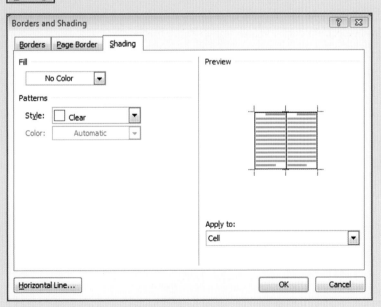

Click on the drop-down arrow on the Fill box and choose the following colour from the palette of colours by clicking on it:

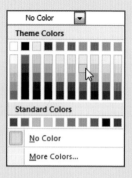

It is important to note that pastel colours are best as background colours to cells in tables if those cells also contain text.

Click on OK to confirm the colour choice and the colour is applied to the column headings like this:

Speed (km/h)	Thinking distance (m)	Braking distance (m)	Overall stopping distance (m)
32	6	6	12
48	9	14	23
64	12	24	36
80	15	38	53
96	18	55	73
112	21	75	96

3 Now shade the other cells until the table looks the same as this:

Speed (km/h)	Thinking distance (m)	Braking distance (m)	Overall stopping distance (m)
32	6	6	12
48	9	14	23
64	12	24	36
80	15	38	53
96	18	55	73
112	21	75	96

4 You are now going to remove the gridlines from the table.

Select the entire table by highlighting it.

Right click and select "Borders and Shading" from the menu.

You need to change the settings on the dialogue box to the following:

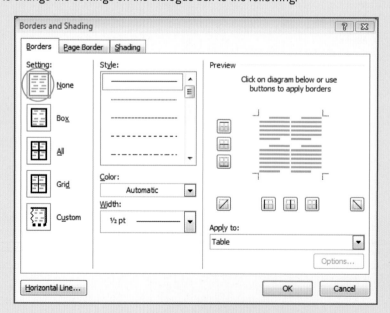

Ensure that "None" has been selected from the Setting and then click on OK.

Also as you do not want any of the borders including the borders to the cells shown, check that "Table" has been selected in the Apply to section. Click on OK to confirm these selections.

Notice that the table now appears like this with dotted lines showing where the borders had been.

Speed (km/h)	Thinking distance (m)	Braking distance (m)	Overall stopping distance (m)
32	6	6	12
48	9	14	23
64	12	24	36
80	15	38	53
96	18	55	73
112	21	75	96

These dotted lines do not appear when the table is printed out. You can print preview the document to see how it will appear when printed.

Speed (km/h)	Thinking distance (m)	Braking distance (m)	Overall stopping distance (m)
32	6	6	12
48	9	14	23
64	12	24	36
80	15	38	53
96	18	55	73
112	21	75	96

5 Save your document using the filename **Stopping_distances_final**

Ensure the accuracy of the text

When producing any document containing text such as a letter, email, presentation, DTP-produced material, etc., it is essential that the text is checked for accuracy. It can be very embarrassing to give a presentation to a large group of people that contains spelling mistakes. Luckily there are software tools to help you such as spellcheckers and grammar checkers.

Using software tools

Spellcheckers

Nearly all document-producing software contains a dictionary against which all the words in a document may be compared to check their spelling. Most allow you to add words to the dictionary, which is particularly useful if you use special terms such as are used in law or medicine. It is important to note that spellchecking a document will not get rid of all the errors. For instance, if you intended to type "the" and typed "he" instead, then the spellchecker will not detect this since "he" as a word is spelt correctly. After using a spellchecker it is still necessary to visually check and/or proof-read a document.

Spell checking

Any document you produce should be spellchecked.

To check the spelling of a document you need to follow these steps:

1 With the document open, click on the Review tab Review and then on  Spelling & Grammar.

2 The following dialogue box appears.

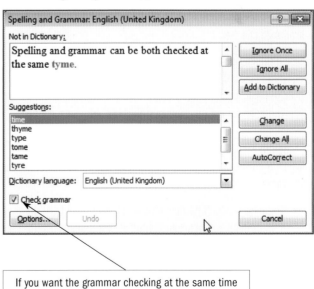

If you want the grammar checking at the same time as the spelling then ensure there is a tick in this box.

You need to look carefully at any word shown in red. This word "tyme" is not in the dictionary because it is spelt incorrectly or mistyped. If the correct word is in the suggestions you can click on the word and then click on Change.

Remember that just because a word is not in the dictionary does not make it incorrectly spelled. There are many words that are specialist words. Such words can be added to the dictionary by clicking on Add to Dictionary so that they will not be queried again in other documents.

Caution

Do not add words to the dictionary unless they are specific to you or your college/school. Don't just add them because you think that is how they are spelled.

Grammar checkers

There are rules about the construction of sentences, and grammar checkers are software tools most often found with word-processing software that will check that these rules have been obeyed. Grammar checkers can be used to check that:

▸ sentences end with only one full stop
▸ there is a capital letter at the beginning of a sentence
▸ common errors like writing "you and I" rather than "you and me" have been avoided.

Proof-reading and correcting documents

Here is some text that someone has typed. Can you spot what is wrong?

I no their are mistakes in my spelling but I will use the spell cheque to cheque them.

If you look at this sentence, you will see that the words are all spelt correctly. You will also notice that some of the words are the wrong ones. The spellchecker will not pick this up because there are no spelling mistakes. Using the grammar check will pick up some mistakes such as the word "their". This should read "there".

Type this sentence then use the spellchecker and then the grammar checker.

Spellchecking and grammar checking will not ensure that your document makes sense. It is therefore important to proof-read your document. It is a good idea to get someone else to read through your work, as they might spot mistakes that you have missed.

Things to watch out for when proof-reading

When proof-reading your document you should look out for the following:

▸ Consistent line spacing – check that you have consistently left the same space between blocks of text, paragraphs, between text and diagrams, etc.
▸ Consistent character spacing – the spacing is usually adjusted automatically in word-processed documents. If characters in the same title do not have the same spacing in a document such as a poster it will look odd.
▸ Re-pagination – during editing, text may have to be added or removed and this can throw the pagination out. For example, you could have a page break in the middle of a paragraph. You could start a new topic at the end of the page rather than at the start of the next page.

▸ Removing blank pages – check that there are no blank pages.
▸ Removing widows/orphans – an orphan is the name given to the first line of a paragraph when it is separated from the rest of the paragraph by a page break. The orphan line appears as the last line at the end of one page, with the rest of the paragraph at the top of the next page. This spoils the look and the readability of the text. A widow is when the last line of a paragraph appears printed by itself at the top of a page and again this can ruin the appearance (and readability) of a document. During your proof-reading you need to check for widows and orphans and adjust the text to remove them.
▸ Dealing with tables and lists split over columns or pages – tables should always be complete with the entire table being shown on the one page. During editing, the table may become split over two pages. You need to spot this and adjust the pages so that the table appears on a single page. Lists normally have a brief explanation of what the list is before the bullet points of the list. If the list is split over two pages, it makes it harder to read and understand.

Activity 10.10

Putting text into columns

In this activity you will learn the following skills:

▸ Set the right, left, top, and bottom page margins
▸ Set the page orientation (i.e. landscape or portrait)
▸ Set the number of columns to use
▸ Set each column width
▸ Set the amount of space between columns

1 Load Microsoft Word and locate and then open the document called **Advantages and disadvantages of networking**

2 Click on Page Layout and then on Margins and select Custom Margins and you will see the dialogue box for the page layout appear.

Set the top and bottom margins to 3 centimetres and the left and right margins to 2 centimetres. When done correctly the settings should be the same as this:

Now set the page orientation to landscape by clicking on the icon like this:

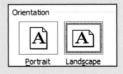

➡

Click on OK to confirm your selections and you will see the document appear on the screen in landscape orientation.

3 Select all the text in the document by highlighting and change it to a sans serif font and change the font size to 14 pt.

4 Select the heading and change the font size of it to 26 pt.

5 Select all the text other than the heading and click on | Page Layout | and then click on | Columns |.
You will be given a choice of how many columns. Choose two from the menu like this:

The text appears like this:

Advantages and disadvantages of networking

Advantages

The ability to share resources (e.g. printers, scanners etc.)

Only one copy of applications software needs to be installed and maintained.

The ability to share data.

Greater security because data and programs may be held in one place.

Everyone using the network is able to access the same centrally held pool of data.

Enables email to be sent between terminals thus saving time and money.

Disadvantages

A network manager/administrator is usually needed and it costs a lot to employ them.

Faults with hardware, communication lines can render the network incapable of being used.

If the network is connected to external communication lines (e.g. the telephone) then there is a danger from hackers.

Viruses, if introduced onto a network, can rapidly spread and cause lots of damage to data and programs.

---

6 Double left click on the blue gap in the centre of the ruler as shown here:

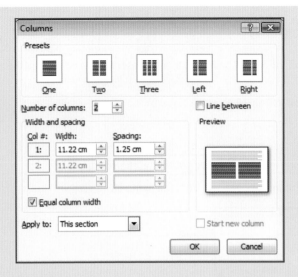

This dialogue box can be used to alter the width of each column, put a line between the columns and alter the spacing between columns.

You are going to alter the spacing between the columns to 1 centimetre.

Alter the spacing to 1 cm like this:

Click on OK to confirm these new settings.

7 Click on the start of "Disadvantages" and press return so that the heading "Disadvantages" appears at the top of the second column.

Your document should now look like this:

Advantages and disadvantages of networking

Advantages

The ability to share resources (e.g. printers, scanners etc.)

Only one copy of applications software needs to be installed and maintained.

The ability to share data.

Greater security because data and programs may be held in one place.

Everyone using the network is able to access the same centrally held pool of data.

Enables email to be sent between terminals thus saving time and money.

Disadvantages

A network manager/administrator is usually needed and it costs a lot to employ them.

Faults with hardware, communication lines can render the network incapable of being used.

If the network is connected to external communication lines (e.g. the telephone) then there is a danger from hackers.

Viruses, if introduced onto a network, can rapidly spread and cause lots of damage to data and programs.

---

8 Create a header and add your name left aligned and the number 121121 right aligned.

9 Create a footer and place an automated page number aligned to the right margin.

10 Save your file as a Word file in your area using the filename **Advantages and disadvantages of networking**

11 Print a copy of your document.

Gutter margin

A gutter margin is used when document is to be bound. Adding a gutter margin tells the computer to add more space so that when the pages are bound, all of the content on the page can be seen.

To add a gutter margin you should perform the following steps:

1   Click on   Page Layout   and select   Margins   and from the menu select Custom Margins.

2   The Page Setup dialogue box appears. Here you can apply the gutter to the left or the top of the document. Notice also that you can adjust the size of the gutter in cm. Here the gutter position is set to the left and the size of the gutter has been set to 1cm.

You would then need to click on OK to apply the gutter to the whole document.

11 Data manipulation

In this chapter you will be learning about how to manipulate data in order to solve problems. You will learn how to create databases using specialist database software and use this software to perform calculations on the data and process the data in different ways. You will also learn how to produce reports in order to output information from the database.

This chapter uses the software MS Access. MS Access is a brand name so you have to make sure that you do not use MS Access in an examination question when the answer is 'database software'. MS Access is popular database software but there are other examples of databases that can be chosen.

The key concepts covered in this chapter are:
▶▶ Creating a database structure
▶▶ Manipulating data
▶▶ Presenting data

Create a database structure

Database structures
You came across an introduction to database structures in Chapter 5. It would be wise to revisit this chapter to make sure you understand the concepts of files, records, and fields. You will also need to understand what is meant by a key field and how data in databases is organized in tables.

You will remember that data is stored in tables with the field names across the first row and the fields containing the data in the columns below. A record is a row of data in the table. The tables are linked to each other, enabling data in any of the tables to be combined if needed. Once data has been entered into the table or tables and any links created, the database can be used.

A query is used to extract specific information from a database, so queries are used to ask questions of databases. For example, a query could be used to extract the names of pupils in a school who are aged 16 years or over. Queries are usually displayed on the screen but they can be printed out if needed. If a printout is needed, it is better to produce a report. A report is a printout of the results from a database. Reports can be printed out in a way that is controlled by the user.

Field types
You also came across field types in Chapter 5. To put data into a database structure you need to decide the field type for each field. Here is a summary of the main field types:

Alphanumeric/text – this is used to hold letters and also numbers that are not to be used in calculations.

Numeric – this field is used for true numbers (i.e., not phone numbers or long code numbers). If it is sensible to use the number in a calculation then a numeric field should be used. Once you have decided that a field is to be numeric you can then set the following:

▶▶ Decimal – you can set the numeric field to hold numbers having a specified number of decimal places.
▶▶ Integer – you can set up the numeric field to Integer (for fewer than 3 digits) or long integer (for 3 or more digits).
▶▶ Percentage – this turns the number into a percentage and adds a percentage sign when displayed. If you enter 10 into a percentage field it will be displayed as 10% and when used in a calculation the number can be used without having to divide it by 100.

Currency – this is used to hold numbers which are to have a currency symbol attached to them (e.g., $240) when displayed.

Date/time – used to hold dates and times. In many cases only the date is needed for dates of birth, etc. There are many different date formats to choose from (e.g., dd/mm/yy – 12/03/11 for 12 March 2011 or mm/dd/yy – 03/12/11 for 12 March 2011).

Logical/Boolean – logical/Boolean data can have only one of two values: True or False. Any data which can be stored as two possibilities such as true/false, 1 or 0, yes or no can be stored as a logical/Boolean data type.

Using meaningful file and field names
Field names need to be chosen carefully as they need to give an indication of the data they are used to hold.

File names should be chosen carefully so that the name indicates the data held and also so that the file can be found easily.

Activity 11.1

Locating, opening, and importing data from an existing file

In this activity you will learn the following skills:

▸▸ How to identify the structure of a csv file

▸▸ How to import a file in csv file format into database software

▸▸ How to change the names of fields and change the field types

▸▸ How to change the structure of a database

1 Load the word-processing software Word and locate and open the file called **Database of employees**. This file is a comma separated variable (csv) file and a section of it is shown here:

```
Forename,Surname,Sex,DOB,No of IGCSEs,Incl
English,Position,Salary,Full or part time,
Yasmin,Singh,F,12/03/1992,3,Y,Y,Web design
Mohamed ,Bugalia,M,01/09/1987,10,Y,Y,Progr
Viveta,Karunakaram,F,09/10/1978,5,Y,Y,Prog
```

Notice that the first row of the data which also appears on the next line because it is long, contains the field names. Each subsequent row represents a record. Notice that all the data is separated by commas.

Rather than type all this data into the database, we can create a database using this existing file.

Close the file without saving it and exit the word-processing software.

2 Load the software Microsoft Access and then click on

Give the database file the name "Employee database" as shown here:

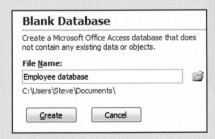

Click on **Create**.

3 Look at the following section of the screen which appears:

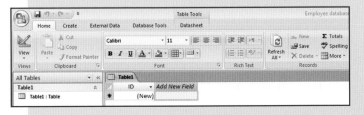

Click on the external data tab **External Data**.

As a csv file is a text file click on **Text File**, which tells the database we want to import the data from a text file.

4 The Get External Data window appears. Click on **Browse...** and find the file called **Database of employees**. Click on it and select Open. The file name appears in the window as shown:

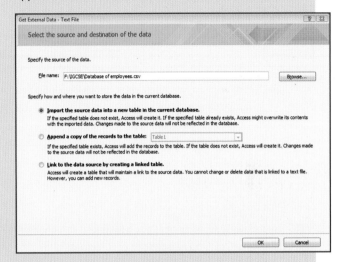

The data needs to be imported into a database table. Check that the first item in the list has been selected and then click on OK.

5 The following window appears:

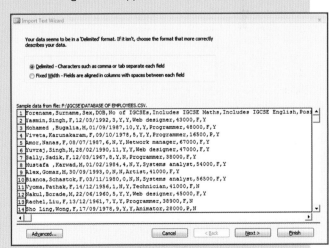

Check that Delimited has been selected as the data being imported has fields separated by commas and then click on **Next >**.

6 The following window appears:

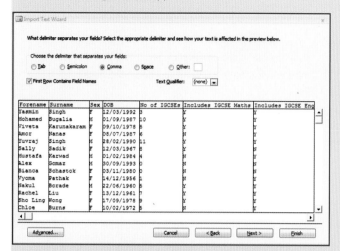

Check that comma has been selected as the delimeter (i.e., what separates the fields in the file) and tick the box for "First Row Contains Field Names".

You are now going to check the data types for each of the fields to make sure they have been set up with the most appropriate data type.

Click on 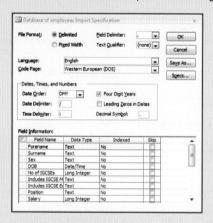 Advanced...

7 The following window appears where you can check the field names that have been set up automatically:

Here you can make changes to the names of the fields and change the data types.

You need to follow the instructions in the examination exactly and make all the changes asked.

Make the following changes:

- ▸▸ Change the data type of the field "Salary" to currency.
- ▸▸ Change the data type of the field "Includes IGCSE Maths" from Text to Yes/No.
- ▸▸ Change the data type of the field "Includes IGCSE English" from Text to Yes/No.

- ▸▸ Change the data type of the field "Driving licence held" from Text to Yes/No.
- ▸▸ Change the field name "Includes IGCSE Maths" to "IGCSE Maths".
- ▸▸ Change the field name "Includes IGCSE English" to "IGCSE English".

After all these changes have been made, click on OK .

8 Notice that the changes have been made in the window shown here:

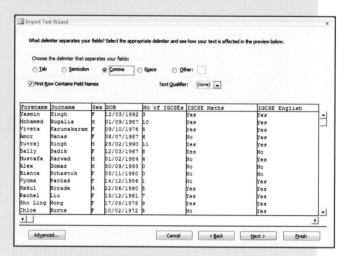

Click on Next > .

The following window appears:

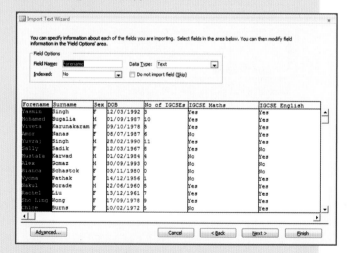

There are no changes to be made here so click Next > .

9 The following window appears:

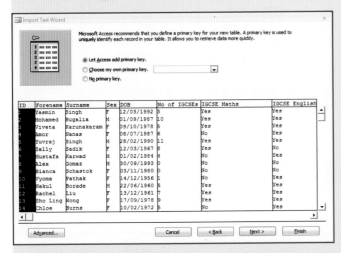

Notice that a new field has appeared in the first column called ID. This is called the Primary Key and it is created automatically to ensure that there is a unique field for each record (i.e., row of the table).

Click on **Next >**.

10 The following window appears:

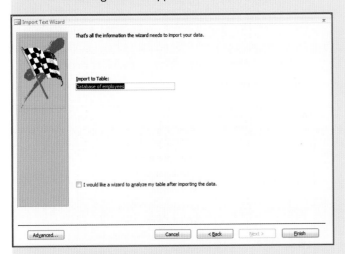

Here you give the table a name. It is a good idea to put the initials "tbl" as the start of the filename and then you know by looking at the name that the file contains a table. Change the name to "tblEmployees" like this:

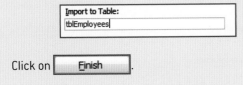

Click on **Finish**.

The following screen appears:

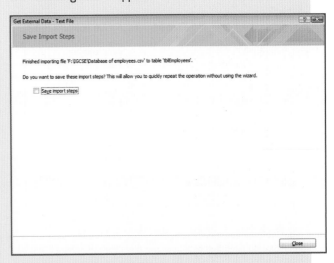

Click on **Close** to complete the import of the file to create a table of a database.

You will see the table listed on the left of the screen like this:

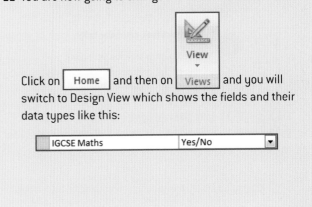

To open the table, double click on it and the table is displayed.

The data is shown in Datasheet view and data can be changed in the table in this view.

Notice that the data for the fields IGCSE Maths, IGCSE English, and Driving licence appears as 0 and -1 so this will need to be altered in the next step.

In order to change the structure of the table you need to change the view to Design View.

11 You are now going to change the structure of the table.

View

Click on **Home** and then on **Views** and you will switch to Design View which shows the fields and their data types like this:

Click on the Data Type for IGCSE Maths like this:

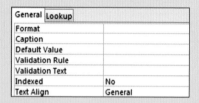

Field Name	Data Type
ID	AutoNumber
Forename	Text
Surname	Text
Sex	Text
DOB	Date/Time
No of IGCSEs	Number
IGCSE Maths	Yes/No
IGCSE English	Yes/No
Position	Text
Salary	Currency
Full or part time	Text
Driving licence held	Yes/No

Now look at the Field Properties section:

General	Lookup
Format	
Caption	
Default Value	
Validation Rule	
Validation Text	
Indexed	No
Text Align	General

Click on the box to the right of "Format" and you will see a drop-down arrow from which you should select Yes/No.

Make similar changes to the following fields:

IGCSE English

Driving licence held.

View

Click on | Views | to view the data after the changes have been made.

You will now see the following box appear:

Microsoft Office Access

You must first save the table.

Do you want to save the table now?

| Yes | No |

Click on Yes to save the changes made and the table containing the data will be displayed. Notice that the Yes/No fields are now showing the data correctly.

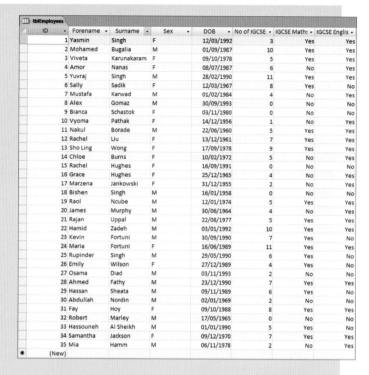

ID	Forename	Surname	Sex	DOB	No of IGCSE	IGCSE Maths	IGCSE Englis
1	Yasmin	Singh	F	12/03/1992	3	Yes	Yes
2	Mohamed	Bugalia	M	01/09/1987	10	Yes	Yes
3	Viveta	Karunakaram	F	09/10/1978	5	Yes	Yes
4	Amor	Nanas	F	08/07/1987	6	No	Yes
5	Yuvraj	Singh	M	28/02/1990	11	Yes	Yes
6	Sally	Sadik	F	12/03/1967	8	Yes	No
7	Mustafa	Karwad	M	01/02/1984	4	No	Yes
8	Alex	Gomaz	M	30/09/1993	0	No	No
9	Bianca	Schastok	F	03/11/1980	0	No	No
10	Vyoma	Pathak	F	14/12/1956	1	No	Yes
11	Nakul	Borade	M	22/06/1960	5	Yes	Yes
12	Rachel	Liu	F	13/12/1961	7	Yes	Yes
13	Sho Ling	Wong	F	17/09/1978	9	Yes	Yes
14	Chloe	Burns	F	10/02/1972	5	No	Yes
15	Rachel	Hughes	F	16/09/1991	0	No	No
16	Grace	Hughes	F	25/12/1965	4	No	Yes
17	Marzena	Jankowski	F	31/12/1955	2	No	Yes
18	Bishen	Singh	M	16/01/1958	0	No	No
19	Raol	Ncube	M	12/01/1974	5	Yes	Yes
20	James	Murphy	M	30/06/1964	4	No	Yes
21	Rajan	Uppal	M	22/08/1977	5	Yes	Yes
22	Hamid	Zadeh	M	03/01/1992	10	Yes	Yes
23	Kevin	Fortuni	M	30/09/1990	7	Yes	No
24	Maria	Fortuni	F	16/06/1989	11	Yes	Yes
25	Rupinder	Singh	M	29/05/1990	6	Yes	No
26	Emily	Wilson	F	27/12/1989	4	Yes	No
27	Osama	Diad	M	03/11/1993	2	No	No
28	Ahmed	Fathy	M	23/12/1990	7	Yes	Yes
29	Hassan	Sheata	M	09/11/1989	6	Yes	No
30	Abdullah	Nordin	M	02/01/1969	2	No	No
31	Fay	Hoy	F	09/10/1988	8	Yes	Yes
32	Robert	Marley	M	17/05/1965	0	No	No
33	Hassouneh	Al Sheikh	M	01/01/1990	5	Yes	No
34	Samantha	Jackson	F	09/12/1970	7	Yes	Yes
35	Mia	Hamm	M	06/11/1978	2	No	Yes
*	(New)						

You can now close the database by clicking on and selecting Close Database.

If you want to exit the database software, click on and select Exit Access

Table Tools

	Recent Documents
New	1 Employee database
Open	2 Database3
	3 Database2
Save	4 Database1
	5 F:\Part One\DrivingSchool5
Save As ▶	6 F:\Part One\DrivingSchool6
	7 Northwind SQL Project File
Print ▶	8 Northwind SQL Project File
	9 Northwind SQL Project File
Manage ▶	
E-mail	
Publish ▶	
Close Database	

Access Options | ✕ Exit Access

Manipulate data

Activity 11.2

Entering data with 100% accuracy and sorting data

In this activity you will learn the following skills:

» Enter and amend data in a database with 100% accuracy
» Adjust the columns so that all field names are visible
» Perform searches using a single criterion
» Perform sorts

Data can be entered straight into a table and you may be asked to add data to existing data in the examination. The data you add can be added at the end of the table and must be entered with 100% accuracy. This means you must carefully proof-read what you have typed by comparing it with what is written on the examination paper. If you make mistakes in your data entry, it will cost you marks.

1 Load the database software Microsoft Access and open the database file called **Employee database**.

 Double click on the table tblEmployees.

2 You will notice that some of the field names and also some of the data items are not fully visible:

| DOB | No of IGCSE | IGCSE Maths | IGCSE Englis | Position | Salary (US $ | Full or part t | Driving licer |

 There is a quick way to adjust this. Click on the white triangle that appears at the top left of the table like this:

 You will notice that the entire table is highlighted. Now move the cursor on the line which separates the field names. You will see it change to a double headed arrow.

 Double left click and the column width will adjust for all the fields so that all the field names and data are shown in the table.

3 At the end of the table enter the following new records. Do not worry about the ID field, as this is inserted automatically.

4 You are now going to use a query to select certain fields to show and also put them in descending order of salary (i.e., with the largest salary first).

Click on the tab 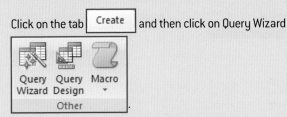 and then click on Query Wizard

The first window of the Query Wizard appears:

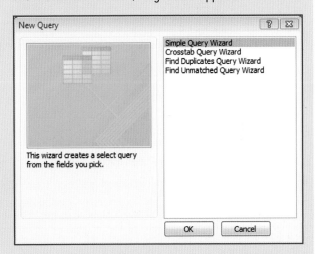

Check that Simple Query Wizard has been selected and click on OK. The next window appears. This is where you select a table and the fields in that table you want displayed in the query. There is only one table in this database.

Forename	Surname	Sex	DOB	No of IGCSEs	IGCSE Maths	IGCSE English	Position	Salary	Full or part time	Driving licence held
Praba	Singh	F	12/01/1994	12	Yes	Yes	Programmer	30000	F	No
Michael	Hazat	M	01/03/1995	6	No	Yes	Programmer	19000	P	Yes
Kevin	Doyle	M	10/09/1992	1	No	No	Web designer	17000	P	Yes
Anesha	Singh	F	30/06/1990	7	Yes	No	Finance clerk	23000	F	No

Proof-read the records you have added.

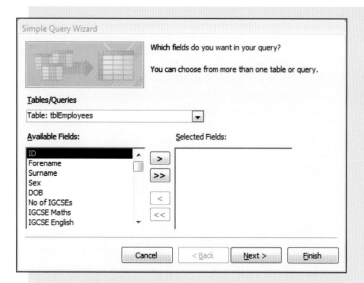

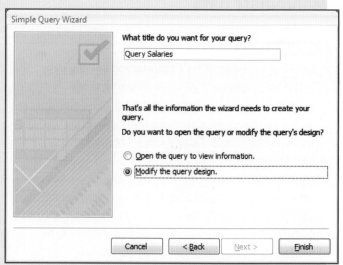

Add the fields by clicking on each of the following fields followed by 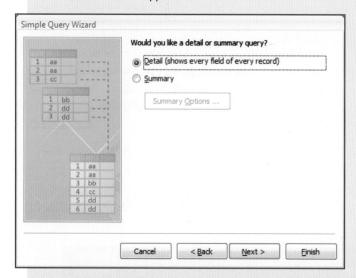 **>** :

Surname
Forename
Salary
Full or part time.

When all these fields have been added, click on **Next >** .

The next window appears:

Change the title for the query to Query Salaries.

Click on the button for Modify the query design and click on **Finish** .

5 The query design is now shown:

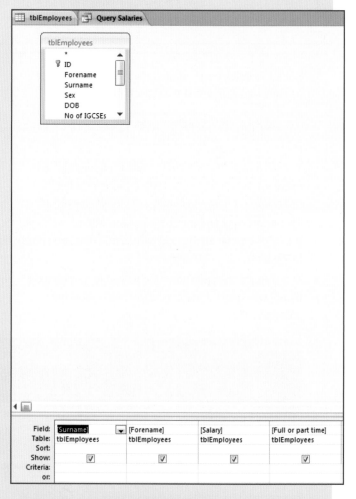

There is nothing to change here so click on **Next >** .

Move the cursor to the Sort box for the Salary field and click on the drop-down arrow and select Descending. This sorts all the data in order of salary with the highest salary first.

Click on to view the results of the query:

Surname	Forename	Salary	Full or part time
Ncube	Raol	£87,000.00	F
Hughes	Grace	£78,000.00	F
Nanas	Amor	£67,000.00	F
Schastok	Bianca	£56,500.00	F
Karwad	Mustafa	£54,000.00	F
Burns	Chloe	£52,000.00	F
Bugalia	Mohamed	£48,000.00	F
Singh	Yuvraj	£47,000.00	F
Borade	Nakul	£45,000.00	F
Singh	Yasmin	£43,000.00	F
Sheata	Hassan	£43,000.00	F
Nordin	Abdullah	£42,000.00	F
Fathy	Ahmed	£41,800.00	F
Pathak	Vyoma	£41,000.00	F
Gomaz	Alex	£41,000.00	F
Singh	Bishen	£39,500.00	F
Liu	Rachel	£38,900.00	F
Sadik	Sally	£38,000.00	F
Wilson	Emily	£37,000.00	F
Hughes	Rachel	£34,200.00	P
Jankowski	Marzena	£32,000.00	P
Jackson	Samantha	£30,500.00	F
Singh	Praba	£30,000.00	F
Wong	Sho Ling	£28,000.00	P
Al Sheikh	Hassouneh	£27,600.00	F
Fortuni	Maria	£27,000.00	F
Murphy	James	£26,000.00	P
Fortuni	Kevin	£26,000.00	F
Singh	Rupinder	£24,900.00	P
Singh	Anesha	£23,000.00	F
Zadeh	Hamid	£23,000.00	F
Diad	Osama	£23,000.00	P
Hoy	Fay	£21,300.00	P
Marley	Robert	£21,000.00	P
Hazat	Michael	£19,000.00	P
Doyle	Kevin	£17,000.00	P
Karunakaram	Viveta	£16,500.00	P
Uppal	Rajan	£14,000.00	P
Hamm	Mia	£12,000.00	P

Record: 1 of 39   No Filter   Search

Click on and then on to save the query.

Activity 11.3

Performing searches using queries

In this activity you will learn the following skills:

▸▸ Perform searches using numeric data with operators (e.g., =, <, >, >=, and <=)

1 If you do not have the **Employee database** already loaded, load the database software Microsoft Access and open the database file called **Employee database**.

Double click on the table tblEmployees.

2 Click on the tab  and then click on Query

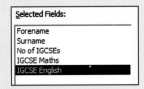

Wizard 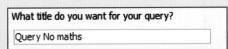.

You will now see the series of windows for the Query Wizard.

Here is what you need to select for each one.

Simple Query Wizard and then click on OK.

Select the following fields and then click on Next:

Selected Fields:
Forename
Surname
No of IGCSEs
IGCSE Maths
IGCSE English

Select ⊙ Detail (shows every field of every record) and then on Next.

3 Enter the following name of the query:

What title do you want for your query?
Query No maths

The purpose of this query is to give information about those employees who do not have IGCSE Maths, so this query name describes the purpose of the query.

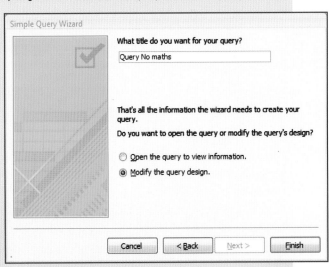

Select 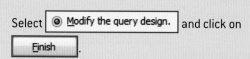 and click on

 .

4 In the Criteria box for the field IGCSE Maths enter No:

Field:	[Forename]	[Surname]	[No of IGCSEs]	[IGCSE Maths]
Table:	tblEmployees	tblEmployees	tblEmployees	tblEmployees
Sort:				
Show:	✓	✓	✓	✓
Criteria:				No

5 To display the results of query click on View.

tblEmployees	tblEmployees QueryNoMaths			
Forename ▾	Surname ▾	No of IGCSEs ▾	IGCSE Maths ▾	IGCSE English ▾
Yasmin	Singh	3	No	Yes
Amor	Nanas	6	No	Yes
Mustafa	Karwad	4	No	Yes
Alex	Gomaz	0	No	No
Bianca	Schastok	0	No	No
Vyoma	Pathak	1	No	Yes
Chloe	Burns	5	No	Yes
Rachel	Hughes	0	No	No
Grace	Hughes	4	No	Yes
Marzena	Jankowski	2	No	Yes
Bishen	Singh	0	No	No
James	Murphy	4	No	Yes
Osama	Diad	2	No	No
Abdullah	Nordin	2	No	No
Robert	Marley	0	No	No
Mia	Hamm	2	No	Yes
Michael	Hazat	6	No	Yes
Kevin	Doyle	1	No	No
*				

6 To close a query click on the cross at the top right of the query screen.

tblEmployees	tblEmployees QueryNoMaths	×

If you are asked to save your query click on Yes.

7 You now have to follow similar steps to the above to create a new query that must display the following fields:

  Forename
  Surname
  Position
  Salary

Give the query the name shown here:

Query High pay

The purpose of the query is to show those fields outlined above for all the employees who have a salary of over 40000 and also to organize the data in descending order (i.e., with the highest salary first).

When you get to the query design screen you need to insert the details in the criteria for the Salary field box >40000 and also click on the sort box and select Descending using the drop down arrow.

Field:	[Forename]	[Surname]	[Position]	[Salary]
Table:	tblEmployees	tblEmployees	tblEmployees	tblEmployees
Sort:				Descending
Show:	✓	✓	✓	✓
Criteria:				>40000

Click on View to see the results and check that they are correct.

Close this query.

Performing searches using wildcards
Here is the data for the field Position:

Position ▾
Web designer
Programmer
Systems analyst
Artist
Systems analyst
Technician
Web designer
Programmer
Animator
Security analyst
Network administrator
Director
Admin clerk
Assistant network manager
Director
Admin clerk
Web designer
Trainee analyst
Trainee analyst
Technician
Finance clerk
Finance clerk
Trainee network engineer
Network engineer
Network engineer
Web designer
Receptionist
Marketing administrator
Marketing administrator
Programmer
Finance clerk
Programmer
Programmer
Web designer
Finance clerk

Suppose you wanted to find any staff with admin in their title. You can see that there are admin clerks and marketing administrators.

In the search criteria box for Position you would need to perform a wildcard search. This means you are looking for any data for this field containing the word "admin".

To perform the wildcard search you would enter: *admin* in the criteria box for the Position field in a query.

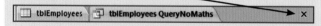

Activity 11.4

Performing wildcard searches

1 Load Microsoft Access and load the file **Employee database** if not already loaded.

2 Create a query containing all the fields. When you have done this the design will look the same as this:

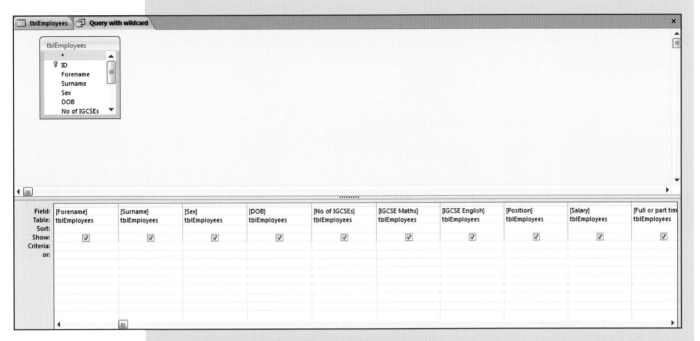

3 In the Criteria box for the field Position enter the following wildcard:

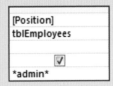

When you enter the formula it changes to the following:

Change the view to Datasheet View and the results are shown.

4 You are now going to alter this query to show only those employees with "admin" in the title for their position AND with a salary of £26,000 or less.

Click on the criteria for Salary and enter <=26000 like this:

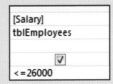

Note that even though this field is formatted to currency you should not include the currency sign with the number in the search criteria.

Notice also that because two criteria are being used, the data has to meet both criteria (i.e., contain the word "admin" for Position AND have a salary of 26000 or less).

Change the view to Datasheet View and the results are shown.

5 Go back to Design View and delete the all the search criteria.

6 You have been asked to show all the details for those employees who are not programmers.

For the Position field enter the criterion Not programmer like this:

```
[Position]
tblEmployees

        ☑
Not programmer
```

When you press enter it changes to the following:

```
Not "programmer"
```

7 Change the view to Datasheet View and the results are shown.

8 Close the query without saving.

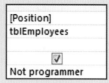

Activity 11.5

Putting formulae in queries

In this activity you will learn the following skills:

▶▶ Putting formulae in queries

1 Load Microsoft Access and load the file **Employee database** if not already loaded.

2 Create a query containing all the fields.

3 Ensure you are looking at the Design View of the query like this:

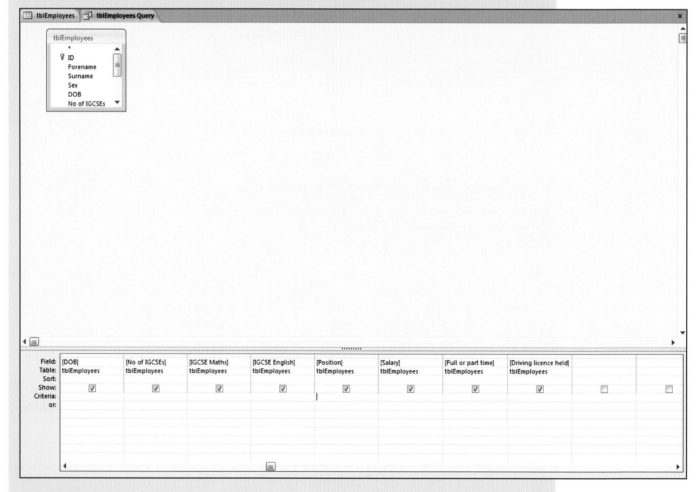

The company has a pension scheme where each employee must pay 8% of their salary into the scheme.

A formula is needed to calculate 8% of the Salary for each employee.

Before entering the formula you should widen the column where the formula is to go so that all the formula can be seen.

In the first blank field enter the following formula:

Pension:([Salary]*8/100)

After entering the formula make sure that the box is ticked to show the results of the calculation.

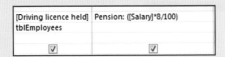

Important notes about this formula:

▸▸ Pension is the name of the new calculated field.

▸▸ Any fields used in calculations must be surrounded by square brackets.

4 Click on Datasheet View to see the results of the query. There is a problem because the data calculated looks like this:

Pension ▾
3440
3840
1320
5360

The calculated Pension field needs to be formatted to currency.

Go back to the Design View and click on the calculated field and in the Property Sheet section click on the drop down arrow for Format and change it to Currency like this:

General	Lookup	
Description		
Format	Currency	▾
Decimal Places		

Now change Decimal Places to 0.

5 Go back to the Datasheet View to see the results of the query.

6 Save this query using the filename Query showing pension.

Activity 11.6

Performing searches on the Employees database

In this activity you will practise the following skills:

▸▸ Perform searches using multiple criteria
▸▸ Use Boolean operators in searches
▸▸ Sort data using one criterion into ascending or descending order

For this activity you are required to produce queries on the Employees database to extract the following data:

1 A list of the fields: surname, forename, sex, and position for all female programmers.

2 A list of all the fields for employees who work part time.

3 A list of the fields: surname, forename, salary, and sex for all employees earning less than 20000.

4 A list of the fields: surname, forename, and position for all employees who are either web designers or programmers.

5 A list of the fields: surname, forename, position, and number of IGCSEs for all the employees having less than or equal to 3 IGCSEs.

6 A list of the fields: surname, position, and salary for all employees who are not programmers.

7 A list of the fields: surname, forename, and position where the position contains the word clerk.

8 A list of the fields: surname, position, and salary for all employees who have the words clerk or analyst as part of the position field.

9 A list of the fields: surname, position, and salary for all employees earning 40000 or over arranged in alphabetical order according to surname.

10 A list of the fields: forename, surname, and position for all female workers who earn over 30000. This list needs to be sorted in descending order according to salary.

Present data

Producing reports

In this activity you will learn the following skills:

» Produce reports to display all the required data and labels in full

» Set report titles, use headers and footers, align data and labels appropriately

A database report is the printout of the output from a database. Reports are professional-looking documents and you have more control over their appearance rather than simply printing out query results.

1 If it is not already loaded, load Microsoft Access and open the file **Employee database**.

2 Click on [Create] and then look for the Reports section shown here:

Click on [Report Wizard].

3 The first window of the Report Wizard appears: This report is going to be based on one of the queries you have already produced and saved.

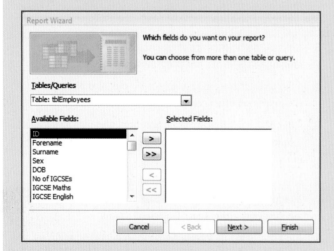

Use the drop-down arrow to find the query:

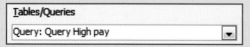

The window now shows the fields that are used in that query like this:

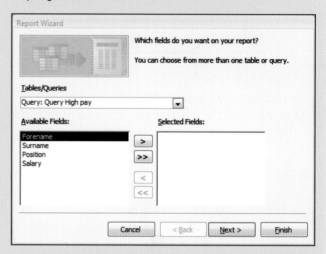

Click on [>>] to add all the fields in one go and click on [Next >].

4 The next screen appears:

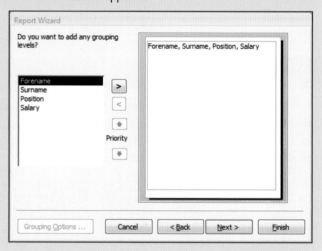

Grouping is not needed for IGCSE so just skip past this by clicking [Next >].

5 The following step asks if you want to sort the field or fields. This is not necessary for this report.

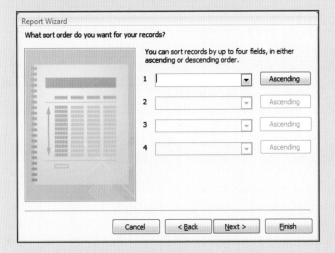

Click on Next >.

6 The following window appears:

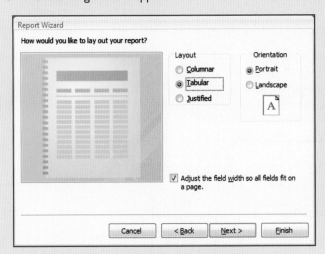

You will be told in the examination paper which orientation to use. Here you will need to change the orientation to Landscape. In the Layout section check that Tabular has been selected like this:

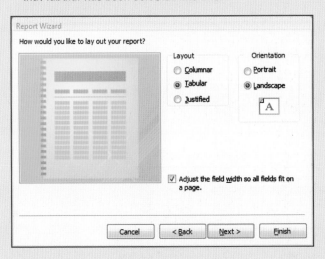

Click on Next >.

7 The next window appears where you can change the style of the report:

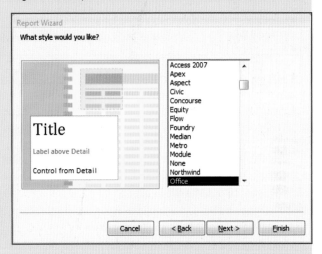

Select Office if it has already not been selected and click on Next >.

8 The next window appears where you need to change the title of the report to "Details of all employees earning over £40000". Also ensure that Modify the report's design has been selected:

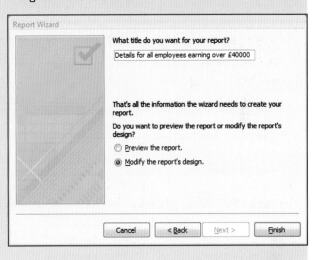

Click on Finish.

9 The design of the report now appears:

Report Header																				
Details for all employees earning over £40000																				

Page Header																				
Forename	Surname					Position							Salary							

Detail																				
Forename	Surname					Position					Salary									

Page Footer																				
=Now()																="Page " & [Page] & " of " & [Pages]				

Report Footer

Notice the way the report is divided into sections:

▸▸ Report header – this is used for a title and any other information for the whole report as it only appears once at the start of the document. You have to obey the examination instructions as to what to put into the header.

▸▸ Page header – this appears at the top of each page and is used to hold the field names.

▸▸ Detail – this shows the rows of data.

▸▸ Page footer – this is used to hold details such as the date the report was produced, who produced it, page numbers, etc. Always obey the examination instructions carefully as to what to put into the footer.

You have been asked to enter your own name on the right in the footer.

Room needs to be made in the Page Footer section for this. To create this room move the cursor onto the light blue bar below the Page Footer section and you will see it change to a double headed arrow like that shown here:

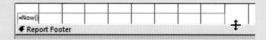

Press the left mouse button down and keeping it pressed down, drag down so that it now looks like this:

Page Footer																				
=Now()																="Page " & [Page] & " of " & [Pages]				

Report Footer

Check that [Design] is selected and then look at the controls section:

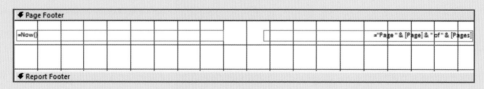

You need to create a Label to contain your name.

Left click on [Label] and move the cursor to the position shown here:

Page Footer																				
=Now()																="Page " & [Page] & " of " & [Pages]				
							+A													

Report Footer

Keeping the left mouse button pressed down, drag the cursor to create the label to look like this:

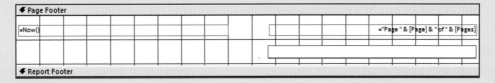

Enter your own name into the label and press Enter:

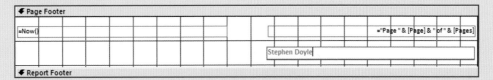

The name needs to be right aligned. To do this click the right align button ⊟.

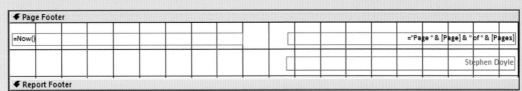

Click on **Views** to view the report and the report is shown like this:

Details for all employees earning over £40000

Forename	Surname	Position	Salary
Raol	Ncube	Director	£87,000.00
Grace	Hughes	Director	£78,000.00
Amor	Nanas	Network manager	£67,000.00
Bianca	Schastok	Systems analyst	£56,500.00
Mustafa	Karwad	Systems analyst	£54,000.00
Chloe	Burns	Security analyst	£52,000.00
Mohamec	Bugalia	Programmer	£48,000.00
Yuvraj	Singh	Web designer	£47,000.00
Nakul	Borade	Web designer	£45,000.00
Hassan	Sheata	Network engineer	£43,000.00
Yasmin	Singh	Web designer	£43,000.00
Abdullah	Nordin	Web designer	£42,000.00
Ahmed	Fathy	Network engineer	£41,800.00
Vyoma	Pathak	Technician	£41,000.00
Alex	Gomaz	Artist	£41,000.00

You can see from the report that the data in the columns for Forename and Surname need more space between them and that some of the letters in the forenames are missing. It is important to always check carefully that all the field names and the data in the columns under them are fully shown.

10 Return to the design of the report by clicking on the drop-down arrow `Views` and then clicking on Design View.

Click on the Forename box in the page header section to select it like this:

Click on the right border of the box and keeping the left mouse button pressed down, drag to the right until the box looks like this:

Notice that the forename box in the detail section underneath the field name is adjusted automatically.

Click on `Views` and then on Report View to see how the report now looks.

Notice that the boxes for the date, page number and your name are not aligned with the other boxes on the left and right sides of the report. Click on each one and adjust it so that it looks like this:

11 The title of the report is to be centralized. To do this click inside the box in the report header to select it:

Report Header									
Details for all employees earning over £40000									

Page Header				
Forename	Surname	Position		Salary

Detail				
Forename	Surname	Position		Salary

Page Footer				
=Now()				="Page " & [Page] & " of " & [Pages]
				Stephen Doyle

Report Footer

Line up the left of the box with all the other items on the page and then stretch the box so it is lined up on the right with the other items. When this has been done it will look like this:

Report Header									
Details for all employees earning over £40000									

Page Header				
Forename	Surname	Position		Salary

Detail				
Forename	Surname	Position		Salary

Page Footer				
=Now()				="Page " & [Page] & " of " & [Pages]
				Stephen Doyle

Report Footer

To centralize the text in this box click on ☰ and the heading will now be in a central position like this:

Report Header									
			Details for all employees earning over £40000						

Page Header				
Forename	Surname	Position		Salary

Detail				
Forename	Surname	Position		Salary

Page Footer				
=Now()				="Page " & [Page] & " of " & [Pages]
				Stephen Doyle

Report Footer

12 Click on **Views** to view the report. Your report should look like this:

Details for all employees earning over £40000

Forename	Surname	Position	Salary
Raol	Ncube	Director	£87,000.00
Grace	Hughes	Director	£78,000.00
Amor	Nanas	Network manager	£67,000.00
Bianca	Schastok	Systems analyst	£56,500.00
Mustafa	Karwad	Systems analyst	£54,000.00
Chloe	Burns	Security analyst	£52,000.00
Mohamed	Bugalia	Programmer	£48,000.00
Yuvraj	Singh	Web designer	£47,000.00
Nakul	Borade	Web designer	£45,000.00
Hassan	Sheata	Network engineer	£43,000.00
Yasmin	Singh	Web designer	£43,000.00
Abdullah	Nordin	Web designer	£42,000.00
Ahmed	Fathy	Network engineer	£41,800.00
Vyoma	Pathak	Technician	£41,000.00
Alex	Gomaz	Artist	£41,000.00

Close the report by clicking on the ☒ positioned at the top right of the report and the following appears:

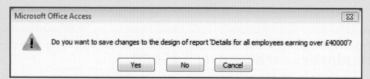

Microsoft Office Access

⚠ Do you want to save changes to the design of report 'Details for all employees earning over £40000'?

[Yes] [No] [Cancel]

Click on Yes and the report will be saved using the filename you entered earlier.

Activity 11.8

Producing calculations at run time

In this activity you will learn the following skills:

▸▸ Create calculated fields
▸▸ Perform calculations at run time
▸▸ Format fields

One way of performing a calculation on data in a database is to create a calculated field which is worked out in a calculation performed on other fields already in the database. This is called a calculation produced at run time.

1 If it is not already loaded, load Microsoft Access and open the file **Employee database**.

2 The company has decided to give all the employees a 5% pay rise. You have been asked to produce a query that shows the fields Forename, Surname, Position, and Salary along with two new calculated fields.

➡

One of these calculated fields is to be called Pay rise and the other is to be called New salary.

Create a query by clicking on Create and then on Query Wizard.

Select Simple Query Wizard and click OK.

Add the fields asked for to the Selected Fields box and then click Next.

At the next window click on Next.

You now change the name of the query to that shown below and also select Modify the query design when your query will look like this:

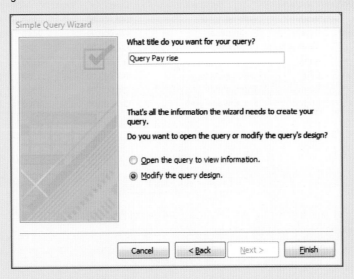

3 Click on Finish and the query design will be shown:

Field:	[Forename]	[Surname]	[Position]	[Salary]
Table:	tblEmployees	tblEmployees	tblEmployees	tblEmployees
Sort:				
Show:	☑	☑	☑	☑
Criteria:				

In the first row of the first blank column enter the following formula:

Pay rise:[Salary]*5/100

There are some important points to remember with formulae:

▸▸ The new field name is written to the left of the colon "Pay rise:".
▸▸ Field names that are used in the formula have to be exactly the same as those used in the database.
▸▸ The field names used in formulae must be enclosed in square brackets.

Adjust the width of the column so that all of this formula can be seen like this:

Field:	[Forename]	[Surname]	[Position]	[Salary]	Pay rise: [Salary]*5/100
Table:	tblEmployees	tblEmployees	tblEmployees	tblEmployees	
Sort:					
Show:	☑	☑	☑	☑	☑
Criteria:					

In the next blank column enter the following formula to calculate the New salary:

New salary:[Salary]+[Pay rise]

Widen the column containing this formula so that all of it is shown like this:

Field:	[Forename]	[Surname]	[Position]	[Salary]	Pay rise: [Salary]*5/100	New salary:[Salary]+[Pay rise] ▾
Table:	tblEmployees	tblEmployees	tblEmployees	tblEmployees		
Sort:						
Show:	✓	✓	✓	✓	✓	✓
Criteria:						
or:						

4 Click on [View] to see the results of the query. Part of the query is shown here:

Forename ▾	Surname ▾	Position ▾	Salary ▾	Pay rise ▾	New salary ▾
Yasmin	Singh	Web designer	£43,000.00	2150	£45,150.00
Mohamed	Bugalia	Programmer	£48,000.00	2400	£50,400.00
Viveta	Karunakaram	Programmer	£16,500.00	825	£17,325.00
Amor	Nanas	Network manager	£67,000.00	3350	£70,350.00
Yuvraj	Singh	Web designer	£47,000.00	2350	£49,350.00
Sally	Sadik	Programmer	£38,000.00	1900	£39,900.00
Mustafa	Karwad	Systems analyst	£54,000.00	2700	£56,700.00

Notice that the Pay rise field is not formatted correctly.

Go back to the Design View and right click the mouse button on the field as shown here:

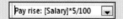

The following menu appears from which you should select Properties like this .

The Property Sheet appears on the right like this:

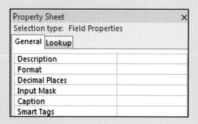

Click on Format and click on the drop down arrow and select Currency from the list.

Click on Decimal Places and click on the drop down arrow and select 2 from the list.

You have now formatted this field.

Just because you see numbers inserted in the correct place do not assume they are correct. It is easy to make mistakes with formulae. Check the calculations manually with a calculator to ensure the output is correct.

5 Click on [View] to see the results of the query. Save and close the query.

Activity 11.9

Creating the report based on the query

In this activity you will learn the following skills:

▸▸ Creating totals of columns of numbers using SUM

1 Load the database software and open the file **Employee database** if it is not already open.

2 Click on Create and then on 🔍 Report Wizard and select the query called "Query Pay rise" created in the last activity. Add all the fields to the Selected Fields area as shown:

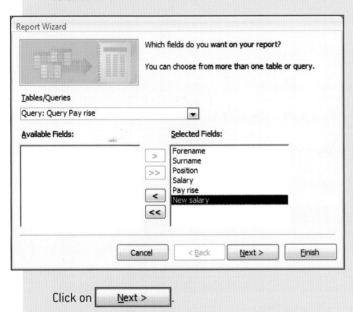

Click on Next > .

3 The next window appears. Just click on Next > .

4 The next window appears where you can sort the data. Sort the Salary field into descending order:

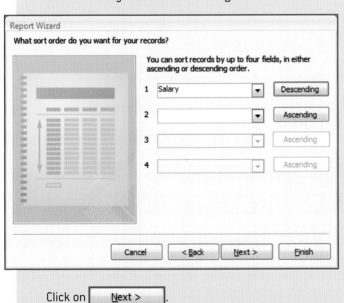

Click on Next > .

5 The next window appears. Make all the changes so it is the same as that shown here:

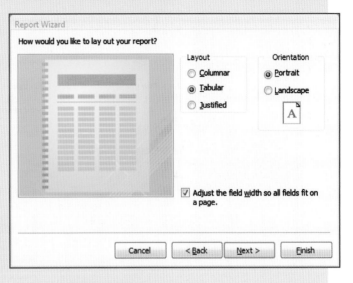

Click on Next > .

6 The following window appears where you can choose a style. Choose the Office style:

Click on Next > .

7 The next window appears where you should change the title to Report Pay rise:

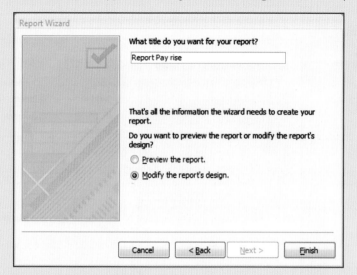

Click on Modify the report's design and then click on | Finish |.

8 Adjust the field names so that all the data is shown when viewed.

9 Stretch the box containing the text "Report Pay rise" so that it fits to the right margin of the report like this:

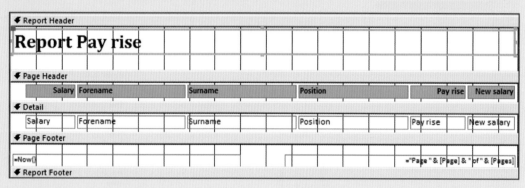

Now stretch the remaining boxes to match the screenshot above.

Change the title of the report in the Report Header to "Report showing a 5% pay rise for all employees" and click on ☰ to centralize it in the box like this:

10 You are going to put some calculations in the Report Footer section of the report.

Click on the horizontal line at the end of the Report Footer section and you will see the double headed arrow. Drag this down until it looks like this:

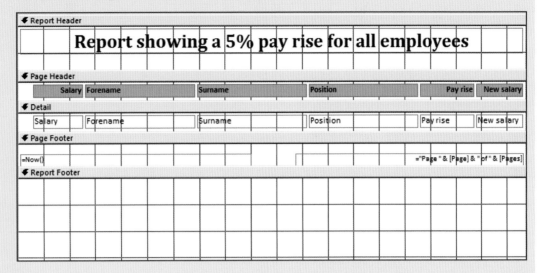

Click on Text Box and drag it to produce a text box below the Salary column.

It needs to look similar to this:

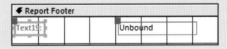

There are actually two boxes here although they are shown on top of each other. One box is used to hold text explaining what the contents of the other box represents.

Click on the small brown box and keeping the left mouse button pressed down, drag to the right until the two boxes appear like this:

◤ Report Footer			
Text19:		Unbound	

Now stretch the text box to the right so that it appears like this:

◤ Report Footer			
Text19:		Unbound	

The box on the right is called a control and this will be used to work out the total of the old salary for all the employees.

Right click on the label (i.e., the box on the left) and you will see a list appear from which you need to select Properties. You will notice the Property Sheet appear down the left-hand side of the screen.

Check that the All tab has been selected in the Property Sheet.

In the Caption box, type the text "Total of old salary":

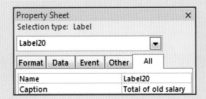

Click on the Controls box (i.e., the box containing the text Unbound) to select it like this:

Report Footer

| Total of old salary | | Unbound | |

Select the Data tab of the Property Sheet like this:

Format	Data	Event	Other	All
Control Source				
Text Format	Plain Text			
Running Sum	No			
Input Mask				
Enabled	Yes			
Smart Tags				

In the Property Sheet section in the box to the right of Control Source enter the following formula:

=SUM([Salary])

Notice the following:

▸▸ The formula starts with an equals sign.

▸▸ The formula is enclosed between two curved brackets.

▸▸ Any fields to which the formula refers must be in square brackets.

You will see the formula inserted in the control box like this:

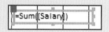

Create two similar text boxes with the following text and formulae:

Total cost of pay rise   =SUM([Pay rise])

Total of new salary   =SUM([New salary])

They should appear like this:

Report Footer

Total of old salary	=Sum([Salary])										
Total cost of pay rise	=Sum([Pay rise])										
Total of new salary	=Sum([New salary])										

11 Click on Report View to see the results of the report. The totals are shown but they are not in currency format.

Click on Design View. If the Property Sheet is not shown on the right then right click on the first control containing the formula and select Properties from the list.

Click on Format and make the following changes

Format	Currency
Decimal Places	2

Repeat this process so that all the controls are formatted to currency.

Click on Report View and you will see all the totals correctly formatted:

Total of old salary	£1,415,700.00
Total cost of pay rise	£70,785.00
Total of new salary	£1,486,485.00

12 Save your report using the filename Report Pay rise.

13 You have been asked to provide evidence of the design of your report.

Click on Design View so that you are looking at the design of your report.

Take a screen shot of the entire screen by pressing the Prt Scr key (you may need to press the shift key first).

Load the word-processing software Microsoft Word and create a new document.

Position the cursor a couple of lines down from the top of the blank document and click on .

You will see the screenshot inserted into the document like this:

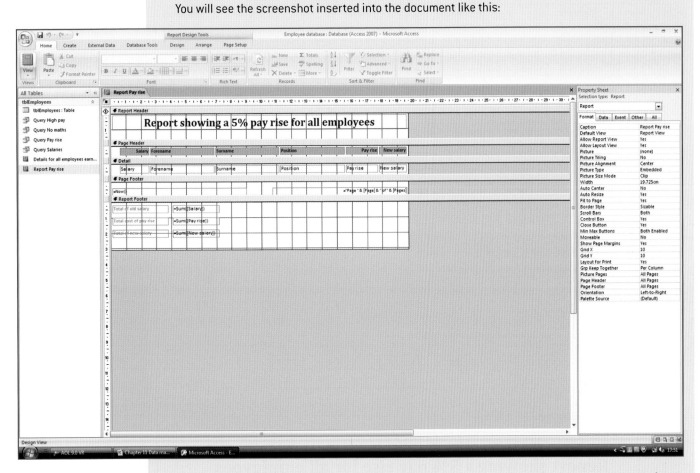

You cannot see the formulae easily so it is necessary to crop the image (i.e., only use part of it) and then enlarge the remaining part.

Left click on the image (i.e., the screenshot) to select it.

Click on [Format] and you will see the picture formatting toolbar like this:

Click on Crop.

Drag the black handles so that only the report design part of the image is shown like this:

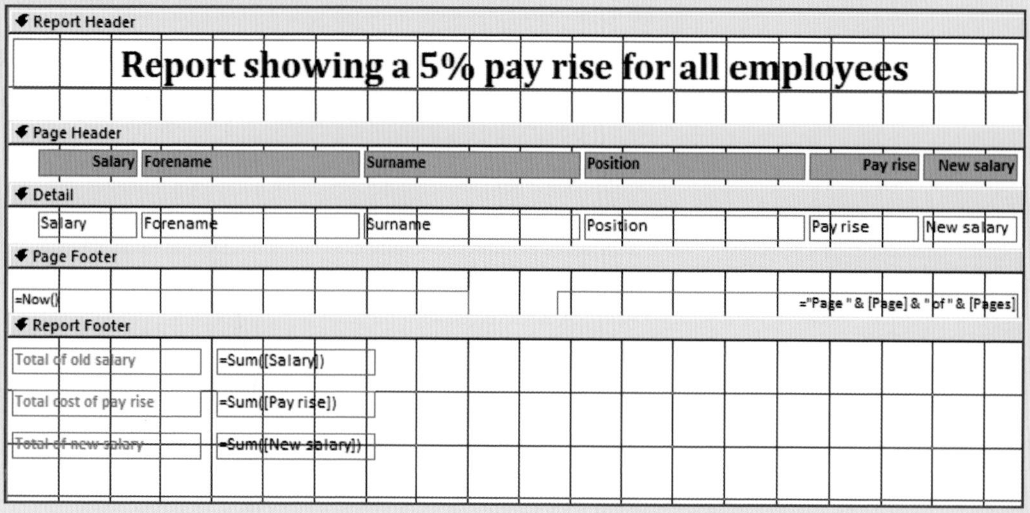

| | Salary | Forename | | Surname | | Position | | Pay rise | New salary |

Report Header

Report showing a 5% pay rise for all employees

Page Header

| Salary | Forename | Surname | Position | Pay rise | New salary |

Detail

| Salary | Forename | Surname | Position | Pay rise | New salary |

Page Footer

=Now()   ="Page " & [Page] & " of " & [Pages]

Report Footer

Total of old salary	=Sum([Salary])
Total cost of pay rise	=Sum([Pay rise])
Total of new salary	=Sum([New salary])

Activity 11.10

Showing and hiding data and labels within a report

In this activity you will learn the following skills:

▶▶ Show and hide data and labels within a report

1   Load Microsoft Access and open the file **Employee database**.

Open the Report Pay rise by clicking on it in the list:

> 🖼 Report Pay rise

2   Change the view to Design view.

You will see the following:

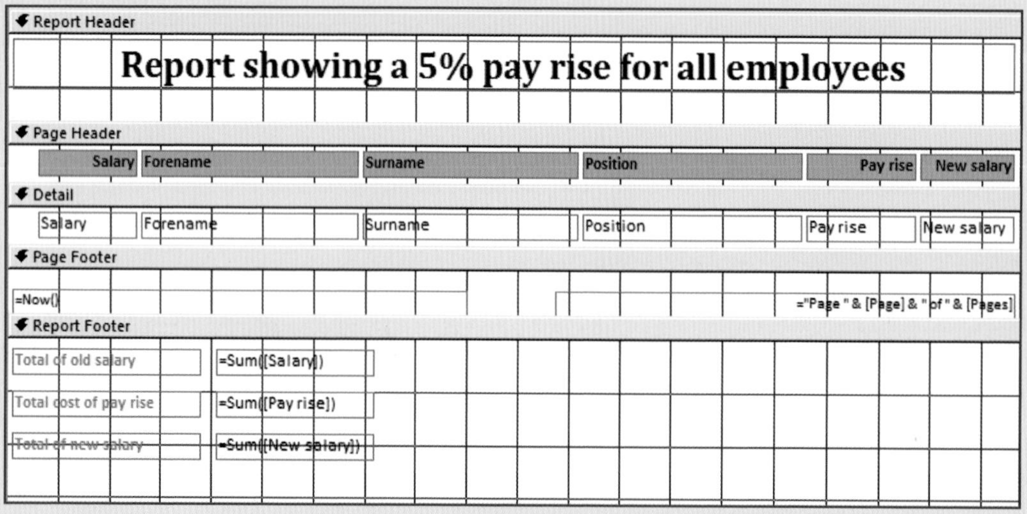

Report Header

Report showing a 5% pay rise for all employees

Page Header

| Salary | Forename | Surname | Position | Pay rise | New salary |

Detail

| Salary | Forename | Surname | Position | Pay rise | New salary |

Page Footer

=Now()   ="Page " & [Page] & " of " & [Pages]

Report Footer

Total of old salary	=Sum([Salary])
Total cost of pay rise	=Sum([Pay rise])
Total of new salary	=Sum([New salary])

3   Sometimes field names, fields or labels need to be hidden.

Here you are going to hide the Salary and New salary details in the above report.

Keeping your finger pressed down on the shift key, left click on the following boxes to select them:

> **Salary**, Salary , **New salary**, and New salary .

➡

154

When they are all selected they will look like this:

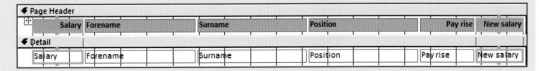

4 Look at the Property Sheet on the right of the screen and select the Format tab if it is not already selected:

Click on the drop-down arrow for the Visible property and select No. This will hide all the items selected.

5 Click on the Report View. The report now appears with the selected items hidden:

Report showing a 5% pay rise for all employees

Forename	Surname	Position	Pay rise
Raol	Ncube	Director	£4,350.00
Grace	Hughes	Director	£3,900.00
Amor	Nanas	Network manager	£3,350.00
Bianca	Schastok	Systems analyst	£2,825.00
Mustafa	Karwad	Systems analyst	£2,700.00
Chloe	Burns	Security analyst	£2,600.00

6 You are now going to show the hidden data.

Go back to the design view where you will notice that there are field names and fields that were not visible in the report. Select all the headings and fields that did not appear in the report and go to the Property Sheet and change Visible to Yes like this:

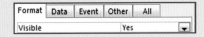

Click on the Datasheet View and the report will now be shown with the previously hidden items displayed.

7 Another way to hide data is to make the background colour the same as the text colour

You are now going to change the background colour to black for the first column of data. We will still keep the column heading and the data below it will look like a black column.

Left click on Salary in the detail section like this:

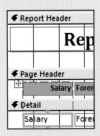

Look at the Property Sheet down the right of the screen and click on <u>Format</u>.

Look down the list until you find the following:

Back Color	#FFFFFF

Left click on the colour and the box will change to:

Back Color	#FFFFFF	▼ …

Click on … and you will see a palette of colours like this:

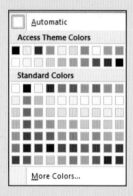

Click on the black colour to set this as the background for the selected field.

8 Go to Report View and you will see the field for Salary hidden like this:

Report showing a 5% pay rise for all employees

Salary	Forename	Surname	Position	Pay rise	New salary
▮	Raol	Ncube	Director	£4,350.00	£91,350.00
▮	Grace	Hughes	Director	£3,900.00	£81,900.00
▮	Amor	Nanas	Network manager	£3,350.00	£70,350.00
▮	Bianca	Schastok	Systems analyst	£2,825.00	£59,325.00
▮	Mustafa	Karwad	Systems analyst	£2,700.00	£56,700.00
▮	Chloe	Burns	Security analyst	£2,600.00	£54,600.00
▮	Mohamed	Bugalia	Programmer	£2,400.00	£50,400.00

9 In a similar way hide the fields for New Salary.

10 Click on  and then click on Save As when the following window appears:

> **Save As** [?] [X]
>
> Save 'Report Pay rise' to:
>
> Copy of Report Pay rise
>
> **As**
>
> Report ▼
>
> [OK] [Cancel]

As the original report has been altered, when the report is saved it is given a different name automatically. Change this name to **Report Pay rise with hidden fields**.

Click on OK.

12 Integration

Integration is important because you frequently have to collect data from different sources and then put it all together in the one document such as a word-processed document or a web page. There are a number of techniques you need to be familiar with when you are integrating data and these are covered in this chapter as well as subsequent chapters.

The key concepts covered in this chapter are:
▸▸ Integrating data from different sources into a single document
▸▸ Importing objects from various sources
▸▸ Combining text with other objects

Integrate data from different sources

Integration is all about using objects created using software or obtained from files, web sites, hardware (such as digital cameras), and then assembling them to form a document. For the examination you will have to create documents that contain many different objects such as spreadsheets, graphs/charts, images etc., so you will have to make sure you can put the components together. You have already come across integration when you transferred data between different software packages.

What is an integrated document?

An integrated document is a document that contains text and other objects such as clip art, photographs, drawings, graphs/charts, spreadsheets, database extracts, etc. Integrated documents include the following:

▸▸ Documents produced using word-processing software – images, graphs/charts, etc., are put together using word-processing software such as Word.
▸▸ Presentations – the slide content may have been obtained from different sources and put together using presentation software such as PowerPoint.

Capturing a screenshot

In the practical examination you may be required to show evidence of how you did something using the software. For example, you might be asked to produce evidence of table design for a database. You might have to provide this in a report (a document outlining your findings when you are asked to do something) and will therefore have to integrate the screenshot with text in a word-processed document.

To produce a screenshot without using specialist software you can:

1 Display the screen you want to capture.
2 If you want a screenshot of everything on the screen, press the Prt Scr key once. If you have more than one window displayed on the screen, then a screenshot of the current window can be obtained by holding down the Alt key and then pressing the Prt Scr key.

3 The screenshot is copied to the temporary storage area called the clipboard. This can be pasted into position in your document, slide, web page, etc.

Using specialist screen capture programs

There are many different screen capture programs and one of the most popular ones is called Snagit. It offers more flexibility than simple screen capture using Ctrl + Alt and then Prt Scr. If you want to try this software then there is a free trial version which you can download onto your computer at: **http://www.techsmith.com/snagit/**

Capturing an image from a web site

You may be asked to locate an appropriate image to be used within a document. Once you have located a suitable image, here are the steps you should take:

1 Display the web page where the image you want to capture is located.
Move the cursor onto the image you want to capture like this:

2 Right click and the following menu appears:

You can either:

▶▶ Click on Save Picture As so that file containing the picture can be saved on disk. You will have to browse to find an appropriate folder in which to store the image and also give the image a name if the one suggested is unsuitable like this:

Or

▶▶ Click on Copy to put a copy of the image into the clipboard. You can then position the cursor on the position where you would like the image to appear in the document. You can then click on Paste to put the image into position.

Importing text from a web site

You may be required to copy text from a web site or email and then incorporate into a document or presentation slide and the steps you need to take are as follows:

1 Display the web page where the text you want to copy is located.
Move to the start of the text and left click and then drag the mouse until the complete section of text has been highlighted like this:

- Underpinning knowledge clearly explained in full-colour double-page spreads providing information in appealing and accessible, bite-sized chunks
- All the knowledge students require written and organised to align precisely with the requirements of the new specification
- Key terminology clearly defined within the text and in a thorough glossary of key terms and phrases
- Up to date, student-friendly business case studies to help students link theory to practice in a meaningful and relevant way
- Activities designed to reinforce key points of learning
- Integrated assessment support with each topic to help boost grades
- Plenty of exam-style practice questions throughout
- Questions with exemplar model answers, mark schemes and examiner commentary
- Visual summary mindmaps for every topic in the specification

2 With the cursor on the highlighted text, right click and select Copy.

3 This puts the selected text into the clipboard.

4 Move to the document that you want the text copied into and click on Paste .

The text will now be copied into the required position. All that you have to do is to ensure that the text (font, font size, font style, font colour, etc.) is consistent with the rest of the document. Check other formatting such as the number of lines after a paragraph, bullets are the same as used in the same document, and so on.

Copying data from a table

If data in a table in a document, slide, web site, etc., needs to be put into a spreadsheet it is easy. You simply select all the required data in the table by highlighting it and then copying it to the clipboard. It can then be pasted into the spreadsheet. You may have to widen columns in order to accommodate the data or headings.

	Maximum	Minimum
Rhodes weather in January	12°C / 54°F	5°C / 41°F
Rhodes weather in February	14°C / 57°F	6°C / 43°F
Rhodes weather in March	16°C / 61°F	7°C / 45°F
Rhodes weather in April	18°C / 64°F	10°C / 50°F
Rhodes weather in May	26°C / 79°F	15°C / 59°F
Rhodes weather in June	29°C / 84°F	17°C / 63°F
Rhodes weather in July	30°C / 86°F	20°C / 68°F
Rhodes weather in August	31°C / 88°F	21°C / 70°F
Rhodes weather in September	27°C / 81°F	18°C / 64°F
Rhodes weather in October	23°C / 73°F	13°C / 55°F
Rhodes weather in November	17°C / 63°F	10°C / 50°F
Rhodes weather in December	14°C / 57°F	7°C / 45°F

Here a table of weather data has been found on the Internet. It has been selected by highlighting it. Right clicking on the data enables the data to be copied into the clipboard.

You can then open the spreadsheet software, position the cursor where you want the data to start and then paste the data into the spreadsheet.

	A	B Maximum	C Minimum	D
1				
2	Rhodes weather in January	12°C 54°F	/ 5°C 41°F	/
3	Rhodes weather in February	14°C 57°F	/ 6°C 43°F	/
4	Rhodes weather in March	16°C 61°F	/ 7°C 45°F	/
5	Rhodes weather in April	18°C 64°F	/ 10°C 50°F	/
6	Rhodes weather in May	26°C 79°F	/ 15°C 59°F	/
7	Rhodes weather in June	29°C 84°F	/ 17°C 63°F	/
8	Rhodes weather in July	30°C 86°F	/ 20°C 68°F	/
9	Rhodes weather in August	31°C 88°F	/ 21°C 70°F	/
10	Rhodes weather in September	27°C 81°F	/ 18°C 64°F	/
11	Rhodes weather in October	23°C 73°F	/ 13°C 55°F	/
12	Rhodes weather in November	17°C 63°F	/ 10°C 50°F	/
13	Rhodes weather in December	14°C 57°F	/ 7°C 45°F	/

There is usually some tidying up to do. For example, here there are some "/" characters that need deleting. If you want to draw graphs, you need to remove the units from the data and put them in the headings instead.

13 Output data

It is important to save your documents using filenames in structured folders so that they can be found quickly when needed. In the examination you will need to supply evidence of the tasks you have completed by including certain printouts on paper.

The key concepts covered in this chapter are:
▸▸ Saving files in different formats
▸▸ Print tables, queries and reports from a database
▸▸ Print emails and attachments

Saving documents

It is possible in the examination that you could forget to do something you have been asked to do. This could mean you would have to start again. To avoid this you should save different versions of your work at various stages. To do this you will need to use filenames that are slightly different, e.g.:

Name	Date modified	Type
Personnel_version 1	09/08/2011 19:48	Word 2007 Docu...
Personnel_version 2	09/08/2011 19:49	Word 2007 Docu...
Personnel_version 3	09/08/2011 19:49	Word 2007 Docu...

If you forgot to do something or need to make an alteration, then it is easy to find an earlier version rather than have to start again.

Make sure you know how to save a file in different file formats such as:

.txt
.csv
.rtf

These are covered in the other chapters.

Saving and printing a draft document

It is important to save your work before you print as problems with the printer might cause the loss of some of your work. Draft documents are documents that you use to check what you have done. Sometimes you have to see how the printed document actually appears on the printed page and it is often easier to proof-read material that has been printed out once it has been spellchecked and grammar checked using the software.

Once a draft document has been checked and any alterations made using the software used to create it, you will need to save the work. You should use a different name for the final document to distinguish it from the draft document.

Draft documents do not need to be in as high a quality as the final document. In order to save expensive printer ink or toner, draft documents are printed in low quality.

Activity 13.1

Printing a draft document
In this activity you will learn the following skills:

▸▸ How to print a draft document

1 Load the software Microsoft Word and the document **Concorde_text**.

2 Click on 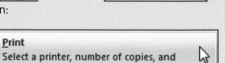 and then on and finally on:

Print
Select a printer, number of copies, and other printing options before printing.

3 The printer dialogue box appears like this:

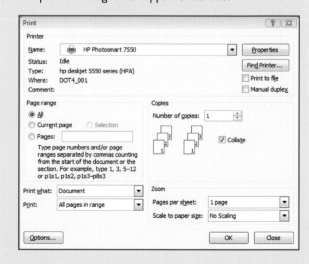

160

Click on Properties .

It now depends on which type of printer you are using what you will see. For an inkjet printer you will see a screen similar to this one:

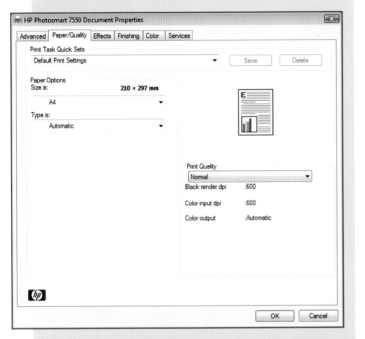

You need to change the Print quality to draft. On this screen you would use the drop-down arrow below Print Quality to change to Draft. With some printers there will be a button that allows you to select the print quality.

Make sure that you know how to produce a draft quality printout using the printer or printers that will be used in the examination room.

4 Print out two copies of the document: one in normal quality and the other in draft quality.

Printing an email

Printing an email is easy as you just click on the print button. The problem is that if you have to provide evidence in the examination that you have attached an email, you will have to take a screenshot of the screen showing the filename of the attachment and the email before you send it. You can then paste the screenshot into a Word document where you can crop the image so only the essential part is shown. You will probably need to re-size the image.

Always check that whatever you have been asked to show appears clearly on the screenshot.

Printing details from Microsoft PowerPoint

In Chapter 16 you will be learning how to produce slides using Microsoft PowerPoint. In the examination you will have to supply evidence that you have successfully completed the instructions given by printing out the following:

▶▶ The slides themselves with the presenter notes.
▶▶ The slides with a certain number to a page with space for the audience to make notes.
▶▶ Screen shots showing animations used (e.g., animations for bullets to appear).
▶▶ Screen shots showing the transitions between slides (i.e., the animation used between one slide and the next).

How you do each of these is covered in Chapter 16.

Printing details from Microsoft Excel

Before you print out any details from a spreadsheet model, you must check carefully that the field names, titles, labels, and data are all visible.

Check that you know how to do the following:

▶▶ Print a copy of the entire model on one or more than one page.
▶▶ Print the results of searches and sorts.
▶▶ Print a copy of the sheet showing the formulae used.

Printing details from Microsoft Access

You may be asked to print items in Microsoft Access such as:

▶▶ Tables
▶▶ Queries
▶▶ Reports.

Printouts of items such as reports, queries and tables are easy to make but you need to check carefully to see if only a section is required or a copy of some aspect of the design such as the calculations use in a report. In these cases you will need to do a screenshot and then copy the screenshot into a word-processor such as MS Word.

If you want to print out a query and have been asked to include your name and candidate number then it is easiest to copy the query (just like you would copy any section of text) and then paste it into a word-processed document. You can then add the details asked for before printing.

Activity 13.2

Printing a query

In this activity you will learn the following skills:

▶▶ How to print a query
▶▶ How to put the results of a query into a word-processed document and add other information

1 Load the software Microsoft Access and the file **Employee database**.

2 Load the query called Query Pay rise.

3 Click on and then on Print and finally on Print Preview like this:

The preview is shown like this:

Notice that the title of the query is given, as are the date and the page number.

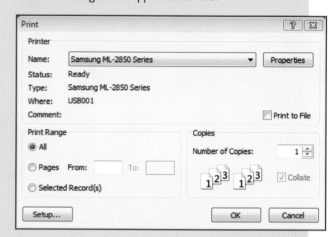

Print this by clicking on [Print].

The Print dialogue box appears like this:

Click on OK and the query is printed.

4 You now have to produce a printout of the query which shows your name and a candidate number printed on the printout. It is necessary to copy the query and paste it into a word-processed document where you can add the name and the candidate number at the top.

Go back to the query being displayed in the Datasheet View.

Click on the ◤ symbol at the top left of the query to select all the data. Once the data is selected it will have a blue background.

Right click anywhere on the data and select Copy from the menu that appears.

5 Load the word-processing software Microsoft Word and create a new document.

Position the cursor three lines down from the top of the document.

Click on [Paste] to paste the query contents into the document like this:

Query Pay rise					
Forename	**Surname**	**Position**	**Salary**	**Pay rise**	**New salary**
Yasmin	Singh	Web designer	£43,000.00	£2,150.00	£45,150.00
Mohamed	Bugalia	Programmer	£48,000.00	£2,400.00	£50,400.00
Viveta	Karunakaram	Programmer	£16,500.00	£825.00	£17,325.00
Amor	Nanas	Network manager	£67,000.00	£3,350.00	£70,350.00
Yuvraj	Singh	Web designer	£47,000.00	£2,350.00	£49,350.00
Sally	Sadik	Programmer	£38,000.00	£1,900.00	£39,900.00
Mustafa	Karwad	Systems analyst	£54,000.00	£2,700.00	£56,700.00
Alex	Gomaz	Artist	£41,000.00	£2,050.00	£43,050.00
Bianca	Schastok	Systems analyst	£56,500.00	£2,825.00	£59,325.00
Vyoma	Pathak	Technician	£41,000.00	£2,050.00	£43,050.00
Nakul	Borade	Web designer	£45,000.00	£2,250.00	£47,250.00
Rachel	Liu	Programmer	£38,900.00	£1,945.00	£40,845.00
Sho Ling	Wong	Animator	£28,000.00	£1,400.00	£29,400.00
Chloe	Burns	Security analyst	£52,000.00	£2,600.00	£54,600.00
Rachel	Hughes	Network administrator	£34,200.00	£1,710.00	£35,910.00
Grace	Hughes	Director	£78,000.00	£3,900.00	£81,900.00
Marzena	Jankowski	Admin clerk	£32,000.00	£1,600.00	£33,600.00
Bishen	Singh	Assistant network manager	£39,500.00	£1,975.00	£41,475.00

6 On the first line enter your name and your candidate number (make this up if necessary).

Left align your name and right align your candidate number. The top of your document will look similar to this:

Stephen Doyle					1212121
Query Pay rise					
Forename	**Surname**	**Position**	**Salary**	**Pay rise**	**New salary**
Yasmin	Singh	Web designer	£43,000.00	£2,150.00	£45,150.00
Mohamed	Bugalia	Programmer	£48,000.00	£2,400.00	£50,400.00
Viveta	Karunakaram	Programmer	£16,500.00	£825.00	£17,325.00
Amor	Nanas	Network manager	£67,000.00	£3,350.00	£70,350.00
Yuvraj	Singh	Web designer	£47,000.00	£2,350.00	£49,350.00

7 Save the document using a suitable filename and print the document out in portrait orientation.

Activity 13.3

Printing a table

In this activity you will learn the following skills:

▸▸ How to print the data in a table
▸▸ How to print out a table design

1 Load the software Microsoft Access and the file **Employee database**.

2 Click on `tblEmployees : Table` to load the table.

3 The table containing the data is shown in Datasheet View. ➡

4 Click on  and then on Print and finally on Print Preview like this:

New	Preview and print the view
Open	**Print** — Select a printer, number of copies, and other printing options before printing.
Save	**Quick Print** — Send the object directly to the default printer without making changes.
Save As ▸	**Print Preview** — Preview and make changes to pages before printing.
Print ▸	
Manage ▸	
E-mail	
Publish ▸	
Close Database	

5 The print view of the table is shown in portrait orientation like this:

	tblEmployees					04/03/2011
ID	**Forename**	**Surname**	**Sex**	**DOB**	**No of IGCSEs**	**IGCSE Maths**
1	Yasmin	Singh	F	12/03/1992	3	Yes
2	Mohamed	Bugalia	M	01/09/1987	10	Yes
3	Viveta	Karunakaram	F	09/10/1978	5	Yes
4	Amor	Nanas	F	08/07/1987	6	No
5	Yuvraj	Singh	M	28/02/1990	11	Yes
6	Sally	Sadik	F	12/03/1967	8	Yes
7	Mustafa	Karwad	M	01/02/1984	4	No
8	Alex	Gomaz	M	30/09/1993	0	No
9	Bianca	Schastok	F	03/11/1980	0	No
10	Vyoma	Pathak	F	14/12/1956	1	No
11	Nakul	Borade	M	22/06/1960	5	Yes
12	Rachel	Liu	F	13/12/1961	7	Yes
13	Sho Ling	Wong	F	17/09/1978	9	Yes
14	Chloe	Burns	F	10/02/1972	5	No
15	Rachel	Hughes	F	16/09/1991	0	No
16	Grace	Hughes	F	25/12/1965	4	No
17	Marzena	Jankowski	F	31/12/1955	2	No
18	Bishen	Singh	M	16/01/1958	0	No
19	Raol	Ncube	M	12/01/1974	5	Yes
20	James	Murphy	M	30/06/1964	4	No
21	Rajan	Uppal	M	22/08/1977	5	Yes
22	Hamid	Zadeh	M	03/01/1992	10	Yes
23	Kevin	Fortuni	M	30/09/1990	7	Yes
24	Maria	Fortuni	F	16/06/1989	11	Yes
25	Rupinder	Singh	M	29/05/1990	6	Yes
26	Emily	Wilson	F	27/12/1989	4	Yes
27	Osama	Diad	M	03/11/1993	2	No
28	Ahmed	Fathy	M	23/12/1990	7	Yes
29	Hassan	Shesta	M	09/11/1989	6	Yes
30	Abdullah	Nordin	M	02/01/1969	2	No
31	Fay	Hoy	F	09/10/1988	8	Yes
32	Robert	Marley	M	17/05/1965	0	No
33	Hassouneh	Al Sheikh	M	01/01/1990	5	Yes
34	Samantha	Jackson	F	09/12/1970	7	Yes
35	Mia	Hamm	M	06/11/1978	2	No
36	Praba	Singh	F	12/01/1994	12	Yes
37	Michael	Hazat	M	01/03/1995	6	No
38	Kevin	Doyle	M	10/09/1992	1	No
39	Anesha	Singh	F	30/06/1990	7	Yes

Click on `Page: 1` ▸ ▸| and you will see that the printout of the whole table takes 3 pages.

➡

Print the table by clicking on 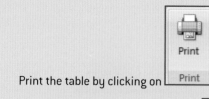.

6 Go to the Design View by clicking on .

7 You have been asked to show evidence of the design.

Take a screenshot of the design and then paste it into a word-processed document. Crop the image and size it so that all the details in the design can be seen clearly.

Add your name and candidate number (you can make this up) at the top of the page with the name left aligned and your candidate number right aligned.

8 Save your document using the filename **Table design**.

9 Print a copy of the document.

Printing reports

Chapter 11 explained how to print database reports. Remember that if you are asked to provide evidence of the design or part of the design such as any formulae used, you will need to take a screenshot and then paste the screenshot into a word-processed document.

Print previewing a report

If you need to print a report out you should print preview it first to check how it will look when printed. This saves paper because you will only be printing your final version. To print out a report you need to display the report on the screen

and then click on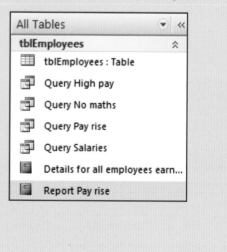

The following menu appears from which you need to select Print and then Print Preview as shown here:

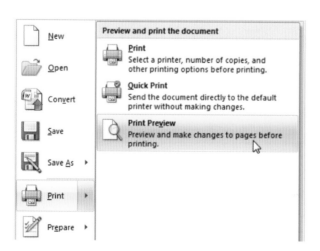

You are then presented with the following toolbar:

Notice that you can select the page orientation: Portrait or Landscape here.

You can also adjust the page margins. This is useful if you need to try to fit all the data on a single page.

Exporting data

In the examination you may be asked to export data from a database by saving it in a file format that can be used by another package. For example, you may want to export data from the database into a word-processed document or a spreadsheet.

You will need to export data in the following common text formats:

▸▸ .csv (comma separated variable file)
▸▸ .txt (text file)
▸▸ .rtf (rich text format file).

To export data from a form, table, query or report you should perform the steps in the next activity.

The following menu appears:

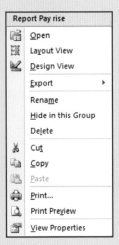

Left click on Export.

The following list of file formats appear from which you should left click on Word RTF file. Notice the other formats in this list (i.e. Text File and Excel).

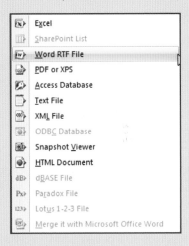

The following window appears where you can select the file name and where the file is to be saved:

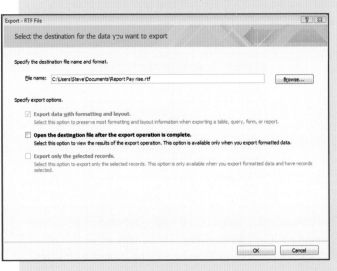

Choose a suitable folder to save your work in and select Open the designation file after the export operation is complete and click on OK.

The data is now appears as a word-processed document like this:

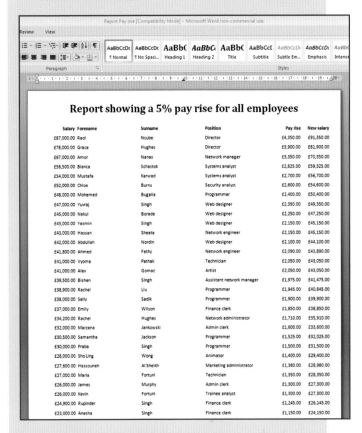

You can now edit the report if you are asked to and then save it as a word-processed file.

If you go back to the database you will see the following window appear:

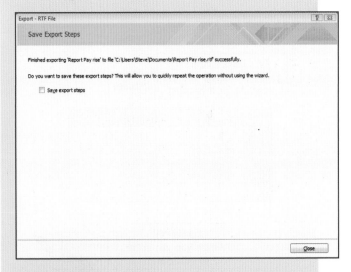

Click on Close.

14 Data analysis

In this chapter you will be learning about spreadsheets and how they are used to create a data model. You will also learn about how such models need to be tested. Once the model has been created, the results from the model can be extracted and presented in a suitable way. For example, you can summarize the results of the model or present the results of the model as a graph or chart. MS Excel has been used here as the spreadsheet software. There are other brands, and other spreadsheet software such as MS Works Spreadsheet and OpenOffice do much the same.

▸▸ The key concepts covered in this chapter are:
▸▸ Creating a data model
▸▸ Testing the data model
▸▸ Manipulating data
▸▸ Presenting data

What is a data model?

A data model is used to mimic a real situation. For example, a data model can be created using spreadsheet software that mimics the money coming into and going out of a business. Data models can be used to provide answers to questions such as "what would happen if I did this?" For example, an economist could look at the effects that raising interest rates would have on the economy.

Creating a data model

In the examination you will be required to produce or adapt a computer model using the spreadsheet skills learnt during your IGCSE course.

Entering the layout of the model

You will be following a set of instructions in the examination and it is important that you enter and position the items that form the model exactly as specified. There are a number of things you will need to know about when laying out a spreadsheet model and these are explained here.

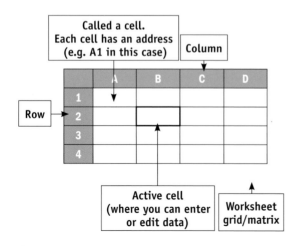

Values

Values are the numbers that are entered into the cells of the spreadsheet.

Labels

Labels are the text next to cells that explain what it is that the cell contains. You should never have a value on a spreadsheet on its own as the user will be left wondering what it represents.

Formulae

Formulae are used to perform calculations on the cell contents. In order to distinguish between text and formulae a symbol, =, needs to be typed in first, like this =B3+B4.

Here are some calculations and what they do. Notice that you can use upper or lower case letters (i.e., capital or small letters):

- = C3+C4 (adds the numbers in cells C3 and C4 together)
- = A1*B4 (multiplies the numbers in cells A1 and B4 together)
- = 3*G4 (multiplies the number in cell G4 by 3)
- = C4/D1 (divides the number in cell C4 by the number in cell D1)
- = 30/100*A2 (finds 30% of the number in cell A2)
- = A2^3 (finds the cube of the number in cell A2)

What must be done with the numbers or contents of cells is determined by the operator. Here is a table of operators and what they do in a formula:

Operator	What it does
+	Add
–	Subtract
*	Multiply
/	Divide
^	Power

Functions

Functions are calculations that the spreadsheet has memorized. For example the function =sum(b3:b10) adds up all the cells from b3 to b10 inclusive. Functions will be covered a bit later.

Copy and paste

Copy and paste can be used to move items around on the same spreadsheet or to copy data from a completely different document or file such as the table in a word-processed document or from a database.

Drag and drop

Drag and drop is a quick way of moving an item. You simply left click the mouse button on the item and then keeping the left mouse button down, drag the item to the new position. When the mouse button is released the item appears in the new position.

AutoFill

Suppose you want to type the days of the week or months of the year down a column or across a row. Excel is able to anticipate what you probably want to do by the first word alone. So, if you type Monday, the chances are that you want Tuesday in the next column or row, and so on. The main advantage in using AutoFill is that the data being entered is less likely to contain errors than if you type in the data yourself. There are many other ways you can use spreadsheet software to fill in data for you, so use the help to find out more about AutoFill.

Manually verifying data entry

In the examination you will be asked to obey a set of instructions. It is extremely important that you obey the instructions exactly. So if you are asked to use a certain font, you must use this even if you feel a different one would be better. Sticking exactly to the requirements is essential to maximize your marks.

Verifying data entry means checking that you have entered the data accurately from the examination paper or other source. You do this carefully by proof-reading. A small mistake in a number in a spreadsheet can change the data in the whole spreadsheet so that it is all incorrect if there are formulae that use the incorrect number.

You will also need to check that you have obeyed all the instructions correctly. It is very easy to miss something out that you should have done.

Mathematical operations and formulae

Spreadsheets can perform all the usual mathematical operations such as:

- Add +
- Subtract –
- Multiply *
- Divide /
- Indices ^ (e.g., square, cube, square root, cube root, etc.)

Activity 14.1

Copying formulae relatively and showing the formulae for a spreadsheet

In this activity you will learn the following skills:

- Enter the layout of a model
- Use mathematical operations
- Copy formulae
- Display formulae

1. Load the spreadsheet software and enter the following data exactly as it is shown here:

	A	B	C	D	E	F
1	Product	Sept	Oct	Nov	Dec	Total
2	Lawn rake	121	56	23	12	
3	Spade (Large)	243	233	298	288	
4	Spade (Medium)	292	272	211	190	
5	Spade (Small)	131	176	149	200	
6	Fork (Large)	208	322	178	129	
7	Fork (Small)	109	106	166	184	
8	Trowel	231	423	311	219	
9	Totals					

2. Enter the function =sum(b2:e2) in cell F2 like this.

	A	B	C	D	E	F
1	Product	Sept	Oct	Nov	Dec	Total
2	Lawn rake	121	56	23	12	=sum(b2:e2)
3	Spade (Large)	243	233	298	288	
4	Spade (Medium)	292	272	211	190	
5	Spade (Small)	131	176	149	200	
6	Fork (Large)	208	322	178	129	
7	Fork (Small)	109	106	166	184	
8	Trowel	231	423	311	219	
9	Totals					

3. Copy the function in cell F2 down the column to cell F8. You do this by moving the cursor to cell F2 containing the formula. Now click on the bottom right-hand corner of the cell and you should get a black cross shape. Hold the left mouse button down and move the mouse down the column until you reach cell F8. You will see a dotted rectangle around the area where the copied formula is to be inserted. Now take your finger off the button and all the results of the calculation will appear. This is called relative copying because the formula is changed slightly to take account of the altered positions of the numbers which are to be added together.

4. Enter the function =sum(b2:b8) into cell B9.

5. Copy the function in cell B9 relatively across the row until cell F9.

Check your spreadsheet looks the same as this:

	A	B	C	D	E	F
1	Product	Sept	Oct	Nov	Dec	Total
2	Lawn rake	121	56	23	12	212
3	Spade (Large)	243	233	298	288	1062
4	Spade (Medium)	292	272	211	190	965
5	Spade (Small)	131	176	149	200	656
6	Fork (Large)	208	322	178	129	837
7	Fork (Small)	109	106	166	184	565
8	Trowel	231	423	311	219	1184
9	Totals	1335	1588	1336	1222	5481

6 You are now going to display the formulae used rather than the values wherever there are calculations.

Click on [Formulas] in the toolbar.

Look at the Formula Auditing section shown here:

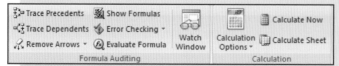

	A	B	C	D	E	F
1	Product	Sept	Oct	Nov	Dec	Total
2	Lawn rake	121	56	23	12	=SUM(B2
3	Spade (Large)	243	233	298	288	=SUM(B3
4	Spade (Medium)	292	272	211	190	=SUM(B4
5	Spade (Small)	131	176	149	200	=SUM(B5
6	Fork (Large)	208	322	178	129	=SUM(B6
7	Fork (Small)	109	106	166	184	=SUM(B7
8	Trowel	231	423	311	219	=SUM(B8
9	Totals	=SUM(B2:B8)	=SUM(C2:C8)	=SUM(D2:D8)	=SUM(E2:E8)	=SUM(F2
10						

Click on Show Formulas.

You should now see the formulae being displayed like this. This is handy as the formulae can be checked.

7 Click on [Show Formulas] again and the spreadsheet will return to showing the values.

8 Save this spreadsheet using the filename "**Product sales**".

Activity 14.2

Setting up a simple model

In this activity you will learn the following skills:

▸▸ Enter data with 100% accuracy
▸▸ Format cells

A university student is living away from home for the first time and they want to make sure that they budget the limited amount of money they will have. They are going to create a spreadsheet model. You are going to follow their steps.

1 Load the spreadsheet software Excel.

2 Enter the details shown on the following spreadsheet exactly as they appear below. Note: you will need to format the text in some of the cells. You will need to format cells D3 and D10 so that the text is wrapped in the cell. (This keeps the cell width the same by moving text so that it fits the width.) You can do this by clicking on the cell where you want the text to be wrapped and then on [Home] and then on [Wrap Text]

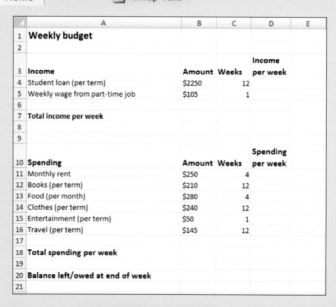

	A	B	C	D	E
1	**Weekly budget**				
2					
3	**Income**			**Income per week**	
4	Student loan (per term)	$2250	12		
5	Weekly wage from part-time job	$105	1		
6					
7	**Total income per week**				
8					
9					
10	**Spending**			**Spending per week**	
11	Monthly rent	$250	4		
12	Books (per term)	$210	12		
13	Food (per month)	$280	4		
14	Clothes (per term)	$240	12		
15	Entertainment (per term)	$50	1		
16	Travel (per term)	$145	12		
17					
18	**Total spending per week**				
19					
20	**Balance left/owed at end of week**				
21					

Note: the columns in row 3 and 10 are labelled "Amount", "Weeks", and "Income per week"/"Spending per week".

3 In cell D4 enter the following formula to work out the amount the student gets per week from the student loan.

=b4/c4

4 In cell D5 enter the following formula to work out the amount of money the student gets each week from their part-time job.

=b5/c5

5 In cell D7 enter the formula =d4+d5 This works out the total weekly income.

6 Enter a formula that will work out the weekly spending on rent and put the answer in cell D11.

7 Copy the formula you have entered in D11 relatively down the column as far as cell D16.

8 Put a formula in cell D18 to add up the spending from cells D11 to D16.

9 Enter a formula in cell D20 which will subtract the total spending from the total income.

10 Your completed spreadsheet will now look like that shown here.

	A	B	C	D	E
1	**Weekly budget**				
2					
3	**Income**	**Amount**	**Weeks**	**Income per week**	
4	Student loan (per term)	$2,250	12	$187.50	
5	Weekly wage from part-time job	$105	1	$105.00	
6					
7	**Total income per week**			$292.50	
8					
9					
10	**Spending**	**Amount**	**Weeks**	**Spending per week**	
11	Monthly rent	$250	4	$62.50	
12	Books (per term)	$210	12	$17.50	
13	Food (per month)	$280	4	$70.00	
14	Clothes (per term)	$240	12	$20.00	
15	Entertainment (per term)	$50	1	$50.00	
16	Travel (per term)	$145	12	$12.08	
17					
18	**Total spending per week**			$232.08	
19					
20	**Balance left/owed at end of week**			$60.42	
21					

11 Save your spreadsheet using the filename "**Budget model**".

Activity 14.3

Using the spreadsheet model to find the answers to "what if" questions

In this activity you will learn the following skills:

▸▸ Amend data in a spreadsheet

▸▸ Print out formulae used

1 Load the spreadsheet called "**Budget model**" if it is not already loaded.

2 You are going to make some changes to the spreadsheet. It is important that you do **not** save any of these changes.

3 The monthly rent has been increased to £275 per month and his employers have reduced his hours for the part-time job, which means he now only earns £50 per week. Will he now be spending more money than he receives? Make these alterations to the spreadsheet to find out. How much does he have left at the end of the week?

4 His grandparents decide they can give him £50 per week to help him. Add this amount to the spreadsheet in a suitable place and using a suitable label. Make any necessary changes to formulae. How much does he now have at the end of the month?

5 Save your spreadsheet using the filename "**Revised budget model**".

6 Print a copy of this spreadsheet on a single page.

7 Print a copy of this spreadsheet showing all the formulae used.

Absolute and relative cell referencing

There are two ways in which you can make a reference to another cell and it is important to know the difference if you want to copy or move cells. When the current cell is copied or moved to a new position, the cell to which the reference is made will also change position.

Look at the spreadsheet below. In relative referencing (which is the normal method of referencing) cell B4 contains a relative reference to cell A1. This reference tells the spreadsheet that the cell to which it refers is 3 cells up and one cell to the left of cell B4. If cell B4 is copied to another position, say E5, then the reference will still be to the same number of cells up and to the left, so the reference will now be to cell D2.

With absolute cell referencing, if cell B4 contains a reference to cell A1, then if the contents of B4 are copied to a new position, then the reference will not be adjusted and it will still refer to cell A1.

In most cases we will want to use relative cell references and the spreadsheet will assume that ordinary cell references are relative cell references. Sometimes we want to refer to the same cell, even when the formula referring to the cell is copied to a new position. We therefore need to make sure that the formula contains an absolute cell reference. To do this, a dollar sign is placed in front of the column and row number.

Cell B6 is a relative cell reference. To change it to an absolute cell reference we would add the dollar signs like this: $B$6. An absolute reference always refers to the same cell whereas a relative reference, refers to a cell that is a certain number of rows and columns away.

Activity 14.4

Absolute and relative cell referencing

In this activity you will learn the following skill:

▸▸ Be able to identify when a relative or absolute cell reference is needed.

1 Load the spreadsheet software and enter the data as shown in the following screenshot.

	A	B	C	D
1	Currency exchange			
2				
3	Current exchange rate		£1 is equivalent to	$1.68
4				
5	Cost of:	Pounds	Dollars	
6	Airline ticket	£299		
7	Hotel	£654		
8	Transfers from airport	£134		
9				

2 The idea of this spreadsheet is to convert the cost of certain items in pounds into dollars.

In order to do this you would put a formula in cell C6 which multiples the cost of the item in cell B6 and multiply it by the conversion factor which is in cell D3.

Enter the formula in cell C6 as shown here:

	A	B	C	D
1	Currency exchange			
2				
3	Current exchange rate		£1 is equivalent to	$1.68
4				
5	Cost of:	Pounds	Dollars	
6	Airline ticket	£299	=D3*B6	
7	Hotel	£654		
8	Transfers from airport	£134		
9				

To save having to type in similar formulae for C7 and C8 you can simply copy the formula from C6 down the column. Try this and you will get the following result

	A	B	C	D
1	Currency exchange			
2				
3	Current exchange rate		£1 is equivalent to	$1.68
4				
5	Cost of:	Pounds	Dollars	
6	Airline ticket	£299	$502.32	
7	Hotel	£654	$0.00	
8	Transfers from airport	£134	$0.00	
9				

You can see that copying the formula did not work.

➡

3 Delete the contents of cells C7 and C8.

4 Now change the formula in cell C6 so that it includes an absolute cell reference to cell D3.

The formula to be entered is: =$D$3*B6.

The formula looks like this in the spreadsheet:

	A	B	C	D
1	Currency exchange			
2				
3	Current exchange rate	£1	is equivalent to	$1.68
4				
5	Cost of:	Pounds	Dollars	
6	Airline ticket	£299	=$D$3*B6	
7	Hotel	£654		
8	Transfers from airport	£134		
9				

5 As cell D3 has an absolute cell reference (i.e. it appears as $D$3 in the formula) the spreadsheet will keep referring to cell D3 even when the formula is moved or copied.

Copy the formula down the column as far as cell C8.

You will now see the formula copied to give the correct results as shown here:

	A	B	C	D
1	Currency exchange			
2				
3	Current exchange rate	£1	is equivalent to	$1.68
4				
5	Cost of:	Pounds	Dollars	
6	Airline ticket	£299	$502.32	
7	Hotel	£654	$1,098.72	
8	Transfers from airport	£134	$225.12	
9				

6 Save the spreadsheet using the filename **Relative and absolute referencing**.

Named cells

Rather than refer to a cell by its cell address (e.g., B3) we can give it a name and then use this name in formulae or whenever else we need to refer to it. For example, if cell B3 contains an interest rate then we can use the name Interest_rate.

Activity 14.5

Naming a cell

In this activity you will learn the following skills:

▶▶ Naming a cell
▶▶ Using named cells in formulae

1 Load Excel and create a new document and input the following data exactly as appears here:

	A	B	C
1	A spreadsheet to work out interest received		
2			
3	Amount of money invested	$20,000	
4	Interest rate	5%	
5	Amount of interest received		
6			
7			

2 Click on cell B4 as this is the cell being given a name.

3 Click on Formulas in the toolbar and then choose Define Name ▾. The following window opens:

Notice that a name has been suggested by using the name in the adjacent cell. You can either keep the name suggested or change it. Here, we will keep the name so click on OK.

Important note: you cannot have blank spaces in a name for a cell. This means rather than use "Interest rate" we use "Interest_rate" as the name.

4 Using what you have learnt from step 3. Give cell B3 the name:

Amount_of_money_invested

5 We will now create the formula for working out the interest in cell B5 using the cell names in the formula rather than the cell addresses.

Click on cell B5 and type in = and click on Formulas (if the formula menus and icons are not shown) and then click on f_x Use in Formula ▾

You will see that the names for the two cells are displayed. Click on Amount_of_money_invested. You will see this is inserted into the formula.

Type in * and then click on f_x Use in Formula ▾ again and this time click on Interest_rate. These two named cells are used in the formula to calculate the interest received.

6 Your final spreadsheet should now look like this:

B5	▾			f_x	=Amount_of_money_invested*Interest_rate		

Activity 4a Named cell

	A	B	C	D	E	F
1	A spreadsheet to work out interest received					
2						
3	Amount of money invested	$20,000				
4	Interest rate	5%				
5	Amount of interest received	1000				

Naming cells makes it easier to understand how the formula works and makes it easier to test.

7 Cell B5 needs to be formatted to currency (dollars with no decimal places).

Do this and you have completed this simple spreadsheet.

Save this spreadsheet using the filename **Activity 5 Named cell**

CSV files

In this file format, which can be identified by the file extension .csv, only text and values are saved. If you save an Excel file in this file format all rows and characters are saved. The columns of data are saved separated by commas, and each row of data in the CSV file corresponds to a row in the spreadsheet. CSV files are used to enable data to be imported or exported between different applications without the need for re-typing if there is no direct way of doing this.

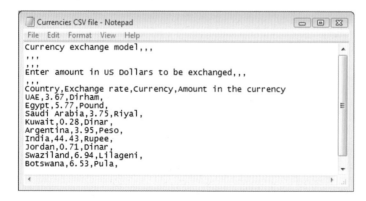

Here is a CSV file that has been created using a non-Microsoft spreadsheet package. It was saved in CSV file format so that it can be loaded into Microsoft Excel.

Activity 14.6

Naming cell ranges

In this activity you will learn the following skills:

▸▸ Importing data in CSV file format
▸▸ Naming a cell range
▸▸ Formatting cells
▸▸ Sorting data

1 Load Excel and open the file in CSV file format called **Currencies CSV file**

Check you have the following displayed on your screen:

	A	B	C	D	E	F
1	Currency exchange model					
2						
3						
4	Enter amount in US Dollars to be exchanged					
5						
6	Country	Exchange	Currency	Amount in the currency		
7	UAE	3.67	Dirham			
8	Egypt	5.77	Pound			
9	Saudi Arab	3.75	Riyal			
10	Kuwait	0.28	Dinar			
11	Argentina	3.95	Peso			
12	India	44.43	Rupee			
13	Jordan	0.71	Dinar			
14	Swaziland	6.94	Lilageni			
15	Botswana	6.53	Pula			

2 When this file was saved in csv format, the formatting was lost, so we need to tidy up this spreadsheet. Format the spreadsheet as follows:

Change the font size of the main heading (i.e., Currency exchange model) to 20 pt and make the text bold.

Make the text "Enter amount in US Dollars to be exchanged" bold.

Widen the columns so that all the column headings and the text in the columns are shown properly.

Make all the column headings bold.

Check that your spreadsheet now looks like this:

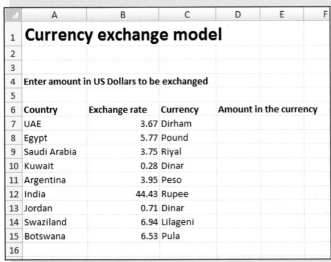

	A	B	C	D	E	F
1	**Currency exchange model**					
2						
3						
4	**Enter amount in US Dollars to be exchanged**					
5						
6	**Country**	**Exchange rate**	**Currency**	**Amount in the currency**		
7	UAE	3.67	Dirham			
8	Egypt	5.77	Pound			
9	Saudi Arabia	3.75	Riyal			
10	Kuwait	0.28	Dinar			
11	Argentina	3.95	Peso			
12	India	44.43	Rupee			
13	Jordan	0.71	Dinar			
14	Swaziland	6.94	Lilageni			
15	Botswana	6.53	Pula			
16						

3 Give cell D4 the following name: Amount_in_Dollars

Look back at the previous activity if you have forgotten how to name a cell.

4 You are now going to name a range of cells.

Click on cell B7 and drag down to cell B15 so that all the cells are highlighted.

Click on 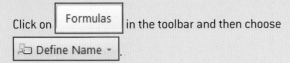 in the toolbar and then choose Define Name ▾.

The following window opens:

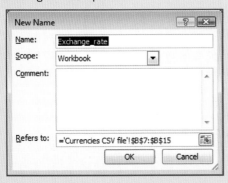

Notice that a name has been suggested and this name is suitable, so just click on OK to give the range of cells this name.

5 Now give the cells in the range D7 to D15 the name

    Amount_when_converted

6 Once the cells have been named they can be used in the formulae, and doing this makes them easier to understand.

In cell D7 enter the following formula:

    =Amount_in_Dollars*Exchange_rate

Notice that you do not use a relative reference for cell D4. The computer knows that this cell must have an absolute reference because the name refers to a single cell and not a range.

7 Enter the number 1 into cell D4. This is the amount in dollars to convert into the other currencies.

8 Now copy the formula in cell D4 down the column as far as cell D15.

	A	B	C	D	E	F
1	**Currency exchange model**					
2						
3						
4	**Enter amount in US Dollars to be exchanged**			1		
5						
6	**Country**	**Exchange rate**	**Currency**	**Amount in the currency**		
7	UAE	3.67	Dirham	3.67		
8	Egypt	5.77	Pound	5.77		
9	Saudi Arabia	3.75	Riyal	3.75		
10	Kuwait	0.28	Dinar	0.28		
11	Argentina	3.95	Peso	3.95		
12	India	44.43	Rupee	44.43		
13	Jordan	0.71	Dinar	0.71		
14	Swaziland	6.94	Lilageni	6.94		
15	Botswana	6.53	Pula	6.53		
16						
17						

9 Change the amount in US Dollars to exchange in cell D4 to 250.

Notice that the values all now change in cells D7 to D15.

Check your spreadsheet looks like this:

	A	B	C	D	E	F
1	**Currency exchange model**					
2						
3						
4	**Enter amount in US Dollars to be exchanged**			250		
5						
6	**Country**	**Exchange rate**	**Currency**	**Amount in the currency**		
7	UAE	3.67	Dirham	917.5		
8	Egypt	5.77	Pound	1442.5		
9	Saudi Arabia	3.75	Riyal	937.5		
10	Kuwait	0.28	Dinar	70		
11	Argentina	3.95	Peso	987.5		
12	India	44.43	Rupee	11107.5		
13	Jordan	0.71	Dinar	177.5		
14	Swaziland	6.94	Lilageni	1735		
15	Botswana	6.53	Pula	1632.5		
16						
17						

10 It is easier to find data in a list if it is sorted into alphabetical order.

The countries need to be put into alphabetical order. To do this, highlight the data in cells A7 to D15.

Important note: If you just highlight the countries these will be put into order but the accompanying data will stay where it is thus jumbling the data up.

Once the data is highlighted click on [Data] in the

toolbar and then on [A-Z Sort].

Click on the ascending order (i.e., A to Z) button [A↓Z].

11 Save the file as an Excel workbook (i.e., not a csv file) using the filename **Currencies**.

Formulae and functions

A function is a calculation that the spreadsheet software has memorized. There are many of these functions, some of which are very specialized. A function must start with an equals sign (=) and it must have a range of cells to which it applies in brackets after it.

Sum: =SUM(E3:P3) displays the total of all the cells from cells E3 to P3 inclusive.

Average: For example, to find the average of the numbers in a range of cells from A3 to A10 you would use: =AVERAGE(A3:A10)

Maximum: =MAX(D3:J3) displays the largest number in all the cells from D3 to J3 inclusive.

Minimum: =MIN(D3:J3) displays the smallest number in all the cells from D3 to J3 inclusive.

Mode: =MODE(A3:A15) displays the mode (i.e., the most frequent number) in the cells from A3 to A15 inclusive.

Median: =MEDIAN(B2:W2) displays the median of the cells from cells B2 to W2 inclusive.

COUNT: Suppose we want to count the number of numeric entries in the range C3 to C30. We can use =COUNT(C3:C30). Any blank lines or text entries in the range will not be counted.

COUNTA: This counts the text, numbers, and blank lines. For example it could be used to calculate the number of people in a list of names like this =COUNTA(C3:C30). You have to be careful when you use this function.

ROUND

The ROUND function rounds a number correct to a number of decimal places that you specify. ROUND is used in the following way:

ROUND(number, number of digits)
where number is the number you want rounded off and number of digits is the number of decimal places.

Here are some examples:

=ROUND(3.56678,2) will return the number 3.57
=ROUND(5.43,1) will return the number 5.4

INTEGER

An integer is a positive or negative whole number or zero. Excel uses the INT function to only give the whole number part of a number. It is important to note that it simply chops off the decimal part of a number just to leave the integer part.

=INT(12.99) will return the integer 12
=INT(0.31022) will return the integer 0

IF

The IF function is called a logical function because it makes the decision to do one of two things based on the value it is testing. The IF function is very useful because you can use it to test a condition, and then choose between two actions based on whether the condition is true or false.

The IF function makes use of something called relational operators. You may have come across these in your mathematics lessons but it is worth going through what they mean.

Relational operators

Symbol	Meaning	Examples
=	equals	5 + 5 = 10
>	greater than	5*3 > 2*3
<	less than	-6 < -1 or 100 < 200
<>	not equal to	"Red" <> "White" or 20/4 <> 6*4
<=	less than or equal to	"Adam" <= "Eve"
>=	greater than or equal to	400 >= 200

Here are some examples of the use of a single IF function:

=IF(B3>=50,"Pass","Fail")

This function tests to see if the number in cell B3 is greater than or equal to 50. If the answer is true, Pass is displayed and if the answer is false, Fail is displayed.

=IF(A2>=500,A2*0.5,A2)

This tests to see if the number in cell A2 is greater than or equal to 500. If true, the number in cell A2 will be multiplied by 0.5 and the answer displayed (i.e., 250 will be displayed). If false, the number in cell A2 will be displayed.

Nested functions

A nested function is a function which is inside another function.

Suppose we want to create a spreadsheet which will tell a teacher whether a student has passed, failed or needs to resit a test. To do this we first need a list of students and their marks.

A formula will need to be created that will do the following:

▸▸ If a student scores 40 marks or less then the message "Fail" appears.

▸▸ If a student scores between (and including) 41 to 50 marks then the message "Resit" appears.

▸▸ If a student scores 51 or more marks then the message "Pass" appears.

Assuming that a mark to be tested is in cell B2, the following formula making use of nested functions will do this:

=IF(B2>=51,"Pass",IF(B2>=41,"Resit","Fail"))

This formula works like this: the mark in cell B2 is tested to see if it is equal to or more than 51 marks, and if it is the message "Pass" appears in the cell where the formula is placed. If this condition is not true then the formula moves to the next IF statement which tests the mark to see if it is greater or equal to 41 in which case a "Resit" message appears. If this is false, the message "Fail" appears.

Activity 14.7

Nested formulae

In this activity you will learn the following skills:

▸▸ Use nested formulae

▸▸ Copy formulae relatively

▸▸ Demonstrate that the model works using test data

1   Load Excel and open the file called **Nested functions**.

2   Check you have the following on your screen:

	A	B	C
1	Name	Mark	Action
2	Rachel Liu	27	
3	Rujav Singh	40	
4	Rcardo Vega	78	
5	Steven Gibbs	54	
6	Marietta Fortuni	89	
7	Vasilios Spanos	79	
8	Mustafa Karwad	25	
9	Amy Hughes	32	
10	Lesley Wong	45	
11	Amor Nanas	10	
12	Mohamed Bugalia	47	
13	Paul Wells	65	
14	Chelsea Dickinson	51	
15	Josuha Mathews	97	

3   Enter the following formula which contains nested functions into cell C2.

=IF(B2>=51,"Pass",IF(B2>=41,"Resit","Fail"))

	A	B	C	D	E	F
1	Name	Mark	Action			
2	Rachel Liu	27	=IF(B2>=51,"Pass",IF(B2>=41,"Resit","Fail"))			
3	Rujav Singh	40				

4   Copy this formula relatively down the column from cell C2 to C15.

5   Check that your spreadsheet looks like this:

	A	B	C
1	Name	Mark	Action
2	Rachel Liu	27	Fail
3	Rujav Singh	40	Fail
4	Rcardo Vega	78	Pass
5	Steven Gibbs	54	Pass
6	Marietta Fortuni	89	Pass
7	Vasilios Spanos	79	Pass
8	Mustafa Karwad	25	Fail
9	Amy Hughes	32	Fail
10	Lesley Wong	45	Resit
11	Amor Nanas	10	Fail
12	Mohamed Bugalia	47	Resit
13	Paul Wells	65	Pass
14	Chelsea Dickinson	51	Pass
15	Josuha Mathews	97	Pass

6   Save your spreadsheet using the filename **Nested formulae**.

7   It is quite hard to construct nested formulae correctly. This means that testing is extremely important. You now have to produce a series of tests that will test that the formula being used is producing the correct results.

You need to check the border values for each message and you need to ensure that you use data which will test these. You now have to create a set of marks that will test the spreadsheet.

Enter these marks by replacing the marks in the previous spreadsheet and check that each message being produced is correct.

Save your spreadsheet using the filename **Checking nested formulae**.

COUNTIF

This is used to count the number of cells with data in them that meet a certain condition.

For example, it could be used to find the total number of males in a group of students by counting the number of entries of "Male".

Activity 14.8

Using the COUNTIF function

In this activity you will learn the following skills:

▸▸ Using COUNTA functions
▸▸ Using COUNTIF functions

1 Load the spreadsheet file **Checking nested formulae** saved in Activity 14.7.

2 In cell A18 enter the text "Number of students"

3 You have been asked to create a formula to display the number of students.
Enter the following formula into cell C18
=COUNTA($C$2:$C$15)
Notice the absolute cell reference in this formula. This is done so that if the formula is copied it will always refer to cells C2 to C15.

4 In cell A19 enter the text "Number of students who passed" and in cell C19 enter the formula
=COUNTIF($C$2:$C$15,"Pass")

5 Enter the text for the labels as shown here:

18	Number of students
19	Number of students who passed
20	Number of students who failed
21	Number of students to resit

6 Add the formulae in cells B19 and C19 to count the number of students who failed and who need to resit. Remember that it is best to use absolute cell references in these formulae.

7 Check that your spreadsheet looks like this:

	A	B	C
1	Name	Mark	Action
2	Rachel Liu	27	Fail
3	Rujav Singh	40	Fail
4	Rcardo Vega	78	Pass
5	Steven Gibbs	54	Pass
6	Marietta Fortuni	89	Pass
7	Vasilios Spanos	79	Pass
8	Mustafa Karwad	25	Fail
9	Amy Hughes	32	Fail
10	Lesley Wong	45	Resit
11	Amor Nanas	10	Fail
12	Mohamed Bugalia	47	Resit
13	Paul Wells	65	Pass
14	Chelsea Dickinson	51	Pass
15	Josuha Mathews	97	Pass
16			
17			
18	Number of students		14
19	Number of students who passed		7
20	Number of students who failed		5
21	Number of students to resit		2

8 Save your spreadsheet using the filename **Spreadsheet using countif**

Lookup functions

Lookup functions search for an item of data in a table and then extract the rest of the information relating to that item. The table of items of data is stored either in the same spreadsheet or a different one. The table can also be a completely different file.

For example, if each different item in a shop has a unique product number, a table of product details can be produced with information such as product number, product description, and product price. If a particular product number is entered, the lookup function will retrieve the other details automatically from the table.

This all sounds a bit complicated but hopefully by completing the following activities you will understand their use.

Activity 14.9

Using VLOOKUP

In this activity you will learn the following skills:

▸▸ Use the VLOOKUP function

▸▸ Format cells

In this activity you will be using a function called VLOOKUP. The idea of this spreadsheet is to type in a product code, and the details corresponding to that code will be looked up by looking for a match for the product code in a vertical column.

1 Load Excel and create a new document.

2 Look carefully at the following spreadsheet and enter the details exactly as they appear here:

	A	B	C
1	A spreadsheet using the VLOOKUP function		
2			
3			
4	Product number		
5	Description		
6	Price		
7			
8			
9			
10			
11	Product Number	Product description	Product price
12			
13	12	Red pen	£0.25
14	13	Black pen	£0.25
15	14	Blue pen	£0.25
16	15	Ruler	£0.75
17	16	Correction fluid	£1.99
18	17	Note pad	£1.25
19	18	Stapler	£7.99
20	19	Staples	£1.20
21	20	Paper clips	£1.55
22	21	String	£1.65
23			

The idea is that the user will type a product number into cell B4, and the computer will look for a match in the vertical column of the table. It will then get the product description and product price and enter them automatically into cells B5 and B6.

3 Enter the product number 15 into cell B4.

A VLOOKUP function needs to be placed in cells B5 and B6. This will use the product number in cell B4 to obtain the rest of the details about the product.

In cell B5 enter the formula =VLOOKUP(B4,A13:C22,2)

This tells the computer that to search for the data vertically in the table from cells A13 to C22 until a match with the product number entered in B4 is found. The "2" at the end means that once a match is found to use the data in the second column.

In cell B6 enter the formula =VLOOKUP(B4,A13:C22,3). Notice the last number in this formula is a 3 because the data to be used is now in the third column.

Format cell B6 to currency with 2 decimal places.

Check that your spreadsheet looks like this:

	A	B	C
1	A spreadsheet using the VLOOKUP function		
2			
3			
4	Product number	15	
5	Description	Ruler	
6	Price	£0.75	
7			
8			
9			
10			
11	Product Number	Product description	Product price
12			
13	12	Red pen	£0.25
14	13	Black pen	£0.25
15	14	Blue pen	£0.25
16	15	Ruler	£0.75
17	16	Correction fluid	£1.99
18	17	Note pad	£1.25
19	18	Stapler	£7.99
20	19	Staples	£1.20
21	20	Paper clips	£1.55
22	21	String	£1.65

4 Test your spreadsheet thoroughly by entering different product numbers in cell B4. As part of the testing you would need to test to see what happens when product numbers that do not exist in the list are entered.

5 Save the spreadsheet using the filename **VLOOKUP**.

Activity 14.10

Using HLOOKUP

In this activity you will learn the following skills:

▸▸ Use the HLOOKUP function

In this exercise you will be using a function called HLOOKUP. This is used when the computer needs to look at the data in the table in the horizontal direction to obtain the match.

The idea of this spreadsheet is for the user to type in pupil names in a class with their forms, and the name of the form teacher automatically appears. The code for the form is looked up horizontally in the table until a match is found.

1 Load Excel, create a new document and type in the following spreadsheet exactly as it appears here.

	A	B	C	D	E	F	G	H
1	A spreadsheet using the HLOOKUP function							
2								
3	Student Name	Form	Form teacher					
4	A Leong	11F						
5	L Rae	11P						
6	F Lam	11R						
7	N Wilkes	11T						
8	S Rousos	11S						
9	F Li	11G						
10	H Patel	11M						
11								
12	Form	11F	11G	11H	11J	11K	11L	11M
13	Form teacher	Mrs Mullen	Mr Patel	Mr Li	Mr Vega	Mrs Hughes	Dr Singh	Mrs Fortuni

Important note: the data that is used to find the list (i.e. in cells B12 to H12) must be arranged in ascending order for the HLOOKUP function to work.

2 Enter the following formula in cell C4:

=HLOOKUP(B4,$B$12:$H$13,2)

This formula works by looking at the contents of cell B4 and then comparing it with the contents in the horizontal set of data in cells B12 to H13 until it finds a match. Once the match is found the computer places the data in the second row (i.e., the form teacher) below the matched data into cell C4.

3 Copy the formula in cell C4 down the column as far as cell C10.

Your spreadsheet now contains all the form teachers for the students like this:

	A	B	C	D	E	F	G	H
1	A spreadsheet using the HLOOKUP function							
2								
3	Student Name	Form	Form teacher					
4	A Leong	11F	Mrs Mullen					
5	L Rae	11M	Mrs Fortuni					
6	F Lam	11L	Dr Singh					
7	N Wilkes	11L	Dr Singh					
8	S Rousos	11K	Mrs Hughes					
9	F Li	11G	Mr Patel					
10	H Patel	11M	Mrs Fortuni					
11								
12	Form	11F	11G	11H	11J	11K	11L	11M
13	Form teacher	Mrs Mullen	Mr Patel	Mr Li	Mr Vega	Mrs Hughes	Dr Singh	Mrs Fortuni

4 Save the spreadsheet using the filename HLOOKUP.

Testing the data model

Models should always be thoroughly tested to make sure that they are producing the correct results. It is easy to make a mistake when constructing a formula. Here are some things to watch out for:

▸▸ Mistakes in the formula (e.g., mistakes in working out a percentage).

▸▸ Incorrect cell references being used.

▸▸ Absolute cell references being used in a formula instead of relative cell references and vice versa.

▸▸ The wrong output being produced when data is filtered, sorted, etc.

▸▸ Not formatting the data in a cell containing a formula.

Always check any calculations performed on data using a calculator and check that all the instructions have been obeyed exactly.

Manipulating data

Spreadsheet software such as MS Excel can be used to produce a flat-file database. This means that all the data is stored in a single file or spreadsheet. In order to produce a database using Excel you need to ensure that the field names appear at the top of each column and that the rows of data are directly below these. You must not leave a blank line between the field names and the data. When creating a simple flat-file database using Excel, the field names are at the top of the column and the data representing a record is a row of the spreadsheet.

Activity 14.11

Searching to select subsets of data

In this activity you will learn the following skills:

▸▸ Perform searches on a set of data

▸▸ Use Boolean operators in searches

1 Load the spreadsheet software Excel and open the file **Database of employees**

Check that the file you are using is the same as this one:

	A	B	C	D	E	F	G	H	I	J	K
						Includes	Includes				Driving
					No of	GCSE	GCSE		Salary	Full or	licence
1	Forename	Surname	Sex	DOB	IGCSEs	Maths	English	Position	(US $)	part time	held
2	Yasmin	Singh	F	12/03/1992	3	Y	Y	Web designer	43000	F	Y
3	Mohamed	Bugalia	M	01/09/1987	10	Y	Y	Programmer	48000	F	Y
4	Viveta	Karunakaram	F	09/10/1978	5	Y	Y	Programmer	16500	P	Y
5	Amor	Nanas	F	08/07/1987	6	N	Y	Network manager	67000	F	Y
6	Yuvraj	Singh	M	28/02/1990	11	Y	Y	Web designer	47000	F	Y
7	Sally	Sadik	F	12/03/1967	8	Y	N	Programmer	38000	F	Y
8	Mustafa	Karwad	M	01/02/1984	4	N	Y	Systems analyst	54000	F	Y
9	Alex	Gomaz	M	30/09/1993	0	N	N	Artist	41000	F	Y
10	Bianca	Schastok	F	03/11/1980	0	N	N	Systems analyst	56500	F	Y
11	Vyoma	Pathak	F	14/12/1956	1	N	Y	Technician	41000	F	N
12	Nakul	Borade	M	22/06/1960	5	Y	Y	Web designer	45000	F	Y
13	Rachel	Liu	F	13/12/1961	7	Y	Y	Programmer	38900	F	N
14	Sho Ling	Wong	F	17/09/1978	9	Y	Y	Animator	28000	P	N
15	Chloe	Burns	F	10/02/1972	5	N	Y	Security analyst	52000	F	Y
16	Rachel	Hughes	F	16/09/1991	0	N	N	Network administrator	34200	P	Y
17	Grace	Hughes	F	25/12/1965	4	N	Y	Director	78000	F	Y
18	Marzena	Jankowski	F	31/12/1955	2	N	Y	Admin clerk	32000	P	Y
19	Bishen	Singh	M	16/01/1958	0	N	N	Assistant network manager	39500	F	N
20	Raol	Ncube	M	12/01/1974	5	Y	Y	Director	87000	F	Y
21	James	Murphy	M	30/06/1964	4	N	Y	Admin clerk	26000	P	N
22	Rajan	Uppal	M	22/08/1977	5	Y	Y	Web designer	14000	P	Y
23	Hamid	Zadeh	M	03/01/1992	10	Y	Y	Trainee analyst	23000	F	Y
24	Kevin	Fortuni	M	30/09/1990	7	Y	N	Trainee analyst	26000	F	Y
25	Maria	Fortuni	F	16/06/1989	11	Y	Y	Technician	27000	F	N
26	Rupinder	Singh	M	29/05/1990	6	Y	N	Finance clerk	24900	P	N
27	Emily	Wilson	F	27/12/1989	4	Y	N	Finance clerk	37000	F	Y
28	Osama	Diad	M	03/11/1993	2	N	N	Trainee network engineer	23000	P	N
29	Ahmed	Fathy	M	23/12/1990	7	Y	Y	Network engineer	41800	F	Y
30	Hassan	Sheata	M	09/11/1989	6	Y	N	Network engineer	43000	F	Y
31	Abdullah	Nordin	M	02/01/1969	2	N	N	Web designer	42000	F	N
32	Fay	Hoy	F	09/10/1988	8	Y	Y	Receptionist	21300	P	N
33	Robert	Marley	M	17/05/1965	0	N	N	Marketing administrator	21000	P	N
34	Hassouneh	Al Sheikh	M	01/01/1990	5	Y	N	Marketing administrator	27600	F	Y
35	Samantha	Jackson	F	09/12/1970	7	Y	Y	Programmer	30500	F	Y
36	Mia	Hamm	M	06/11/1978	2	N	Y	Finance clerk	12000	P	N

Notice the column names which act as the field names and the rows of data which are the records.

2 You are now going to search through this set of data and find a subset of this data that satisfies certain criteria.

You need to select the data you want to use. In this case you are going to use all the data so click on

![selection box icon] at the top left of the spreadsheet as this is a quick way of selecting the entire set of data.

Check that you have the Home tab | Home | selected and if not, click on it.

Click on | Sort & Filter ▾ | in the toolbar and then select Filter. You will notice drop-down arrows appear on each field name.

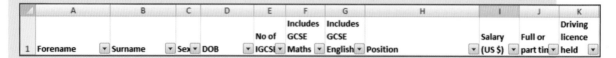

	A	B	C	D	E	F	G	H	I	J	K
						Includes GCSE	Includes GCSE		Salary	Full or	Driving licence
1	Forename ▾	Surname ▾	Sex ▾	DOB ▾	No of IGCSE ▾	Maths ▾	English ▾	Position ▾	(US $) ▾	part tim ▾	held ▾

3 Click on the drop down arrow for Surname and then select Text Filters and finally Equals:

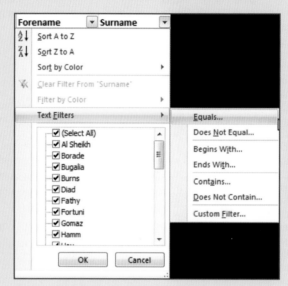

The following window appears where you only view those records that correspond to a certain Surname:

Enter the surname **Singh** into the box like this and then click on OK:

Notice that only the records having the Surname equal to Singh are shown like this:

	A	B	C	D	E	F	G	H	I	J	K
					No of	Includes GCSE	Includes GCSE		Salary	Full or	Driving licence
1	Forename	Surname	Sex	DOB	IGCS	Maths	English	Position	(US $)	part tin	held
2	Yasmin	Singh	F	12/03/1992	3	Y	Y	Web designer	43000	F	Y
6	Yuvraj	Singh	M	28/02/1990	11	Y	Y	Web designer	47000	F	Y
19	Bishen	Singh	M	16/01/1958	0	N	N	Assistant network manager	39500	F	N
26	Rupinder	Singh	M	29/05/1990	6	Y	N	Finance clerk	24900	P	N

4   You are now going to perform a search using multiple criteria.

You need to get back to the complete set of data and to do this you click on the undo button 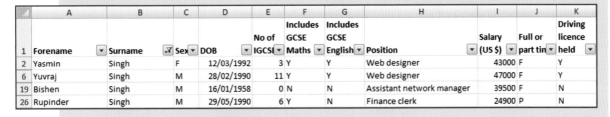.

In this filter you will be displaying only employees having the surname Singh or the surname Hughes.

Click on the drop down arrow for Surname and then select Text Filters and finally Equals.

Now enter the data as shown:

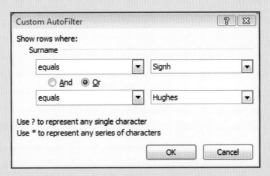

This selects all those records having the surname of either Singh or Hughes.

Click on OK and the data which meets the criteria will be displayed like this:

	A	B	C	D	E	F	G	H	I	J	K
					No of	Includes GCSE	Includes GCSE		Salary	Full or	Driving licence
1	Forename	Surname	Sex	DOB	IGCS	Maths	English	Position	(US $)	part tin	held
2	Yasmin	Singh	F	12/03/1992	3	Y	Y	Web designer	43000	F	Y
6	Yuvraj	Singh	M	28/02/1990	11	Y	Y	Web designer	47000	F	Y
16	Rachel	Hughes	F	16/09/1991	0	N	N	Network administrator	34200	P	N
17	Grace	Hughes	F	25/12/1965	4	N	Y	Director	78000	F	Y
19	Bishen	Singh	M	16/01/1958	0	N	N	Assistant network manager	39500	F	N
26	Rupinder	Singh	M	29/05/1990	6	Y	N	Finance clerk	24900	P	N

5   Go back to the original set of data and ensure that all the data is selected.

Check you have selected all 36 records.

Click on the drop down arrow for Salary and select Number Filters like this:

We want to display all the details for those employees earning 40000 or greater so enter the details as shown into the window:

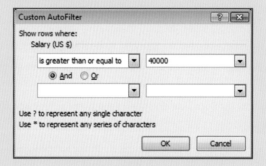

Click on OK and the filtered data will appear like this:

	A	B	C	D	E	F	G	H	I	J	K
	Forename ▼	Surname ▼	Sex ▼	DOB ▼	No of IGCSE ▼	Includes GCSE Maths ▼	Includes GCSE English ▼	Position ▼	Salary (US $) ▼	Full or part tim ▼	Driving licence held ▼
2	Yasmin	Singh	F	12/03/1992	3	Y	Y	Web designer	43000	F	Y
3	Mohamed	Bugalia	M	01/09/1987	10	Y	Y	Programmer	48000	F	Y
5	Amor	Nanas	F	08/07/1987	6	N	Y	Network manager	67000	F	Y
6	Yuvraj	Singh	M	28/02/1990	11	Y	Y	Web designer	47000	F	Y
8	Mustafa	Karwad	M	01/02/1984	4	N	Y	Systems analyst	54000	F	Y
9	Alex	Gomaz	M	30/09/1993	0	N	N	Artist	41000	F	Y
10	Bianca	Schastok	F	03/11/1980	0	N	N	Systems analyst	56500	F	Y
11	Vyoma	Pathak	F	14/12/1956	1	N	Y	Technician	41000	F	N
12	Nakul	Borade	M	22/06/1960	5	Y	Y	Web designer	45000	F	Y
15	Chloe	Burns	F	10/02/1972	5	N	Y	Security analyst	52000	F	Y
17	Grace	Hughes	F	25/12/1965	4	N	Y	Director	78000	F	Y
20	Raol	Ncube	M	12/01/1974	5	Y	Y	Director	87000	F	Y
29	Ahmed	Fathy	M	23/12/1990	7	Y	Y	Network engineer	41800	F	Y
30	Hassan	Sheata	M	09/11/1989	6	Y	N	Network engineer	43000	F	Y
31	Abdullah	Nordin	M	02/01/1969	2	N	N	Web designer	42000	F	N

6 Print out a copy of the filtered data putting the data onto a single page and using landscape orientation.

7 Close the spreadsheet without saving.

Activity 14.12

Searching using wildcards

In this activity you will learn the following skills:

▸▸ Use wildcards in searches

1   Load the spreadsheet software Excel and open the file **Database of employees**

Check that the dataset has 36 records.

2   Select the entire set of data.

3   Click on [Sort & Filter ▾] in the toolbar and then select Filter.

4   Suppose we want to find the subset of data that only contains those surnames having three characters. To do this we use wildcards.

Wildcards are characters that are used as substitutes for other characters in a search.

For example, if we want all surnames with exactly three characters, we use the wildcards **???**

If you wanted all the surnames starting with the letter A you could use the wildcard like this **A***

The difference between the wildcards ? and * are that with ? only a single character is shown but with * any number of characters are shown.

Click on the drop-down arrow for surname and then on Text Filters and finally on Equals.

In the box that appears, enter the wildcard characters **???**:

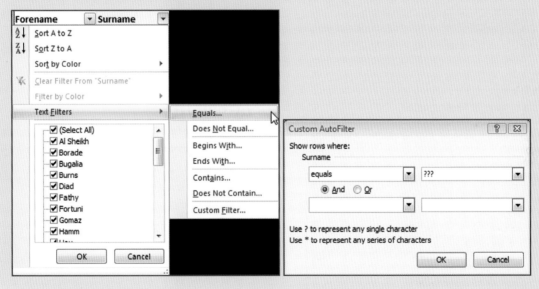

This will pick out the surnames having only any three characters.

Click on OK to display the data:

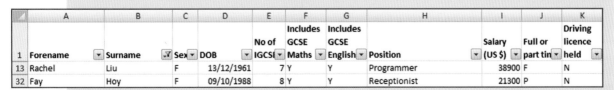

	A	B	C	D	E	F	G	H	I	J	K
					No of	Includes GCSE	Includes GCSE		Salary	Full or	Driving licence
1	Forename	Surname	Sex	DOB	IGCSE	Maths	English	Position	(US $)	part tim	held
13	Rachel	Liu	F	13/12/1961	7	Y	Y	Programmer	38900	F	N
32	Fay	Hoy	F	09/10/1988	8	Y	Y	Receptionist	21300	P	N

5   To remove the filter and get back to the original data you can use Undo but here you will do this another way.

Click on [**Surname** ▾]. Notice the filter symbol showing that data in this column has been filtered.

Click on Clear Filter From Surname.

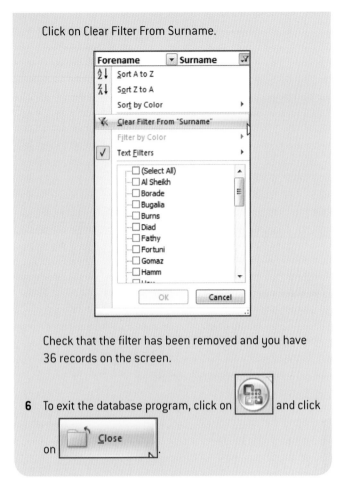

Check that the filter has been removed and you have 36 records on the screen.

6 To exit the database program, click on 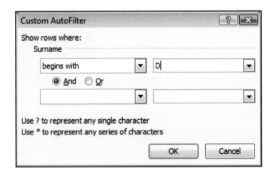 and click on ___.

More about filters

There are lots of ways of filtering data.

This filter shows all the data for employees with surnames starting with the letter "D"

This shows all the details for employees who are either Web designers or Programmers.

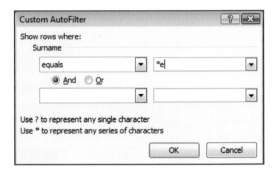

A word about wildcards

To explain how wildcards can be used we will look at how they apply to the field Surname.

This will search for all surnames starting with the letter "S" followed by any 4 characters.

This will search for all surnames which have any combinations of characters providing they end with an "e".

Activity 14.13

Searching for subsets of data in the employee database

In this activity you will practise the following skills:

▶▶ Perform searches using a single criterion and using multiple criteria

▶▶ Perform searches using wildcards

In this activity you need to load the spreadsheet software Excel and the file called **Database of employees** and then perform the following searches by filtering the data. For each search you are asked for, you should produce a printout in landscape orientation and fitted to a single page.

Here are the searches:

1 A list of details for employees who are female.

2 A list of details for employees who earn less than 30 000.

3 A list of all the part-time employees.

4 A list of all the employee details with surnames starting with the letter H.

5 A list of all the employee details with surnames that end with the letter y.

6 A list of all employees who earn less than 50 000 but more than 35 000.

7 A list of all the employees who are either Network engineers or Technicians.

Sorting data

When sorting data you have to select all the data and not simply the field you are performing the sort on. If you do not do this the whole set of data will become jumbled up.

Activity 14.14

Sorting a set of data into ascending or descending order

In this activity you will learn the following skills:

▸▸ Sort data using one criterion

1 Load the spreadsheet software Excel and open the file **Database of employees**

Check that the file you are using is the same as this one:

	A	B	C	D	E	F	G	H	I	J	K
						Includes	Includes				Driving
					No of	GCSE	GCSE		Salary	Full or	licence
1	Forename	Surname	Sex	DOB	IGCSEs	Maths	English	Position	(US $)	part time	held
2	Yasmin	Singh	F	12/03/1992	3	Y	Y	Web designer	43000	F	Y
3	Mohamed	Bugalia	M	01/09/1987	10	Y	Y	Programmer	48000	F	Y
4	Viveta	Karunakaram	F	09/10/1978	5	Y	Y	Programmer	16500	P	Y
5	Amor	Nanas	F	08/07/1987	6	N	Y	Network manager	67000	F	Y
6	Yuvraj	Singh	M	28/02/1990	11	Y	Y	Web designer	47000	F	Y
7	Sally	Sadik	F	12/03/1967	8	Y	N	Programmer	38000	F	Y
8	Mustafa	Karwad	M	01/02/1984	4	N	Y	Systems analyst	54000	F	Y
9	Alex	Gomaz	M	30/09/1993	0	N	N	Artist	41000	F	Y
10	Bianca	Schastok	F	03/11/1980	0	N	N	Systems analyst	56500	F	Y
11	Vyoma	Pathak	F	14/12/1956	1	N	Y	Technician	41000	F	N
12	Nakul	Borade	M	22/06/1960	5	Y	Y	Web designer	45000	F	Y
13	Rachel	Liu	F	13/12/1961	7	Y	Y	Programmer	38900	F	N
14	Sho Ling	Wong	F	17/09/1978	9	Y	Y	Animator	28000	P	N
15	Chloe	Burns	F	10/02/1972	5	N	Y	Security analyst	52000	F	Y
16	Rachel	Hughes	F	16/09/1991	0	N	N	Network administrator	34200	P	N
17	Grace	Hughes	F	25/12/1965	4	N	Y	Director	78000	F	Y
18	Marzena	Jankowski	F	31/12/1955	2	N	Y	Admin clerk	32000	P	Y
19	Bishen	Singh	M	16/01/1958	0	N	N	Assistant network manager	39500	F	N
20	Raol	Ncube	M	12/01/1974	5	Y	Y	Director	87000	F	Y
21	James	Murphy	M	30/06/1964	4	N	Y	Admin clerk	26000	P	N
22	Rajan	Uppal	M	22/08/1977	5	Y	Y	Web designer	14000	P	Y
23	Hamid	Zadeh	M	03/01/1992	10	Y	Y	Trainee analyst	23000	F	Y
24	Kevin	Fortuni	M	30/09/1990	7	Y	N	Trainee analyst	26000	F	Y
25	Maria	Fortuni	F	16/06/1989	11	Y	Y	Technician	27000	F	N
26	Rupinder	Singh	M	29/05/1990	6	Y	N	Finance clerk	24900	P	N
27	Emily	Wilson	F	27/12/1989	4	Y	N	Finance clerk	37000	F	Y
28	Osama	Diad	M	03/11/1993	2	N	N	Trainee network engineer	23000	P	N
29	Ahmed	Fathy	M	23/12/1990	7	Y	Y	Network engineer	41800	F	Y
30	Hassan	Sheata	M	09/11/1989	6	Y	N	Network engineer	43000	F	Y
31	Abdullah	Nordin	M	02/01/1969	2	N	N	Web designer	42000	F	N
32	Fay	Hoy	F	09/10/1988	8	Y	Y	Receptionist	21300	P	N
33	Robert	Marley	M	17/05/1965	0	N	N	Marketing administrator	21000	P	N
34	Hassouneh	Al Sheikh	M	01/01/1990	5	Y	N	Marketing administrator	27600	F	Y
35	Samantha	Jackson	F	09/12/1970	7	Y	Y	Programmer	30500	F	Y
36	Mia	Hamm	M	06/11/1978	2	N	Y	Finance clerk	12000	P	N

2 Click on ▦ at the top left of the spreadsheet to select the entire set of data.

3 Click on  and select Custom sort from the pull-down menu and this Sort box appears:

4 Click on the drop down arrow for Sort by and select the field Surname from the list of fields. Notice that we will be sorting the data in ascending order (A to Z) according to surname:

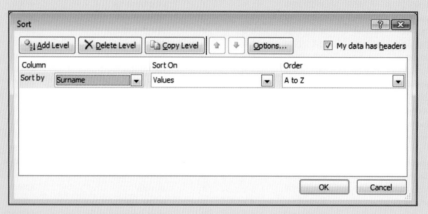

Click on OK to complete the sort.

5 Check that the entire set of data is still selected and sort the data into descending order of Surname. You do this by changing the order of the sort.

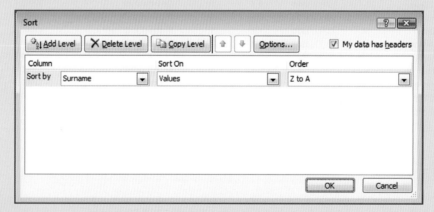

6 Close the file without saving as this will keep the file in its original order.

Activity 14.15

Sorting data using more than one criterion

In this activity you will learn the following skills:

▸▸ Sort data using more than one criterion

1   Load the spreadsheet software Excel and open the file **Database of employees**

2   Select the entire set of data and click on and select Custom sort from the pull-down menu.

You are now going to sort the data into alphabetical order according to position and also for those people in the same position, sort these into numerical order of Salary with the greatest salary first.

Enter the details as shown:

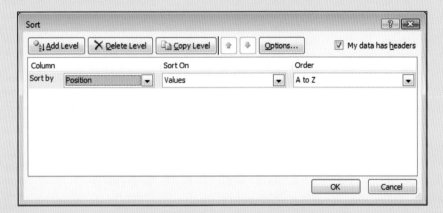

Click on Add Level and in the **Then by** box select the field Salary and finally in the right-hand box select Largest to smallest.

Check your settings are now the same as this:

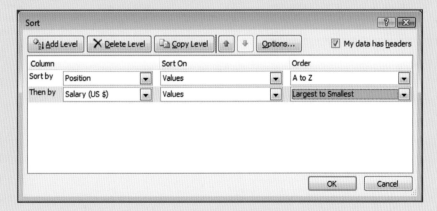

Click on OK to perform the sort.

Check your results are the same as this:

	Forename	Surname	Sex	DOB	No of IGCSEs	Includes GCSE Maths	Includes GCSE English	Position	Salary (US $)	Full or part time	Driving licence held
2	Marzena	Jankowski	F	31/12/1955	2	N	Y	Admin clerk	32000	P	Y
3	James	Murphy	M	30/06/1964	4	N	Y	Admin clerk	26000	P	N
4	Sho Ling	Wong	F	17/09/1978	9	Y	Y	Animator	28000	P	N
5	Alex	Gomaz	M	30/09/1993	0	N	N	Artist	41000	F	Y
6	Bishen	Singh	M	16/01/1958	0	N	N	Assistant network manager	39500	F	N
7	Raol	Ncube	M	12/01/1974	5	Y	Y	Director	87000	F	Y
8	Grace	Hughes	F	25/12/1965	4	N	Y	Director	78000	F	Y
9	Emily	Wilson	F	27/12/1989	4	Y	N	Finance clerk	37000	F	Y
10	Rupinder	Singh	M	29/05/1990	6	Y	N	Finance clerk	24900	P	N
11	Mia	Hamm	M	06/11/1978	2	N	Y	Finance clerk	12000	P	N
12	Hassouneh	Al Sheikh	M	01/01/1990	5	Y	N	Marketing administrator	27600	F	Y
13	Robert	Marley	M	17/05/1965	0	N	N	Marketing administrator	21000	P	N
14	Rachel	Hughes	F	16/09/1991	0	N	N	Network administrator	34200	P	N
15	Hassan	Sheata	M	09/11/1989	6	Y	N	Network engineer	43000	F	Y
16	Ahmed	Fathy	M	23/12/1990	7	Y	Y	Network engineer	41800	F	Y
17	Amor	Nanas	F	08/07/1987	6	N	Y	Network manager	67000	F	Y
18	Mohamed	Bugalia	M	01/09/1987	10	Y	Y	Programmer	48000	F	Y
19	Rachel	Liu	F	13/12/1961	7	Y	Y	Programmer	38900	F	N
20	Sally	Sadik	F	12/03/1967	8	Y	N	Programmer	38000	F	Y
21	Samantha	Jackson	F	09/12/1970	7	Y	Y	Programmer	30500	F	Y
22	Viveta	Karunakaram	F	09/10/1978	5	Y	Y	Programmer	16500	P	Y
23	Fay	Hoy	F	09/10/1988	8	Y	Y	Receptionist	21300	P	N
24	Chloe	Burns	F	10/02/1972	5	N	Y	Security analyst	52000	F	Y
25	Bianca	Schastok	F	03/11/1980	0	N	N	Systems analyst	56500	F	Y
26	Mustafa	Karwad	M	01/02/1984	4	N	Y	Systems analyst	54000	F	Y
27	Vyoma	Pathak	F	14/12/1956	1	N	Y	Technician	41000	F	N
28	Maria	Fortuni	F	16/06/1989	11	Y	Y	Technician	27000	F	N
29	Kevin	Fortuni	M	30/09/1990	7	Y	N	Trainee analyst	26000	F	Y
30	Hamid	Zadeh	M	03/01/1992	10	Y	Y	Trainee analyst	23000	F	Y
31	Osama	Diad	M	03/11/1993	2	N	N	Trainee network engineer	23000	P	N
32	Yuvraj	Singh	M	28/02/1990	11	Y	Y	Web designer	47000	F	Y
33	Nakul	Borade	M	22/06/1960	5	Y	Y	Web designer	45000	F	Y
34	Yasmin	Singh	F	12/03/1992	3	Y	Y	Web designer	43000	F	Y
35	Abdullah	Nordin	M	02/01/1969	2	N	N	Web designer	42000	F	N
36	Rajan	Uppal	M	22/08/1977	5	Y	Y	Web designer	14000	P	Y

3 You now have to follow a series of similar steps to produce the following sort on two criteria:

Sort into numerical order according to salary with the largest first and then into alphabetical order according to surname starting with the letter A first.

4 Produce a printout on a single sheet of paper in landscape orientation.

5 Close the spreadsheet file without saving.

Activity 14.16

Further sorting into two criteria

In this activity you will practise the following skills:

▸▸ Sort data
▸▸ Print output according to instructions

1 Load the spreadsheet **Database of employees**.

2 Sort the entire set of data into Position order according to alphabetical order (A to Z) and then sort into date of birth order with the oldest employees first.

3 Save and print a copy of the sorted data on a single page in landscape orientation.

4 Close the spreadsheet without saving.

Headers and footers

A header is an area between the very top of the page and the top margin. A footer is the area between the very bottom of the page and the bottom margin. Once you tell the spreadsheet you want to use headers and footers then you can insert text into one or both of these areas.

Here are some types of information that is commonly put into the headers and footers:

▸▸ Page numbers
▸▸ Today's date
▸▸ The title
▸▸ A company logo (it can be a graphic image)
▸▸ The author's name
▸▸ The filename of the file that is used to hold the document.

Activity 14.17

Creating a header and footer for the spreadsheet for the database of employees

In this activity you will learn the following skills:

▸▸ Add headers and footers to a spreadsheet
▸▸ Insert information such as date, page numbers, etc., into headers and footers

1 Load the spreadsheet Excel and open the file **Database of employees**.

2 Click on [Insert] and then on Header & Footer.

3 You will see an area at the top of the spreadsheet document like this:

| Click to add header |

Click on this area and a text box appears into which you can enter the details for the header. Enter the text shown here:

| Database of employees last updated by S Doyle |

With the cursor positioned on the line below click on Current Date.

The details in the header will now look the same as this:

| Database of employees last updated by S Doyle &[Date] |

4 You are now going to add the footer.

Click on Footer in the toolbar and a menu appears from which you need to click on **Page 1 of ?**

This will appear in the footer like this:

| Page 1 of 1 |

5 You are now going to see how the header and footer appear if the data was to be printed.

Click on and select Print and then Print Preview.

The spreadsheet containing the header and footer is now shown:

Database of employees last updated by S Doyle
21/09/2011

Forename	Surname	Sex	DOB	No of IGCSEs	Includes GCSE Maths	Includes GCSE English	Position	Salary (US $)	Full or part time	Driving licence held		
Yasmin	Singh	F	12/03/1992	3	Y	Y	Web designer	43000	F	Y		
Mohamed	Bugalia	M	01/09/1987	10	Y	Y	Programmer	48000	F	Y		
Viveta	Karunakaram	F	09/10/1978	5	Y	Y	Programmer	16500	P	Y		
Amor	Nanas	F	08/07/1987	6	N	Y	Network manager	67000	F	Y		
Yuvraj	Singh	M	28/02/1990	11	Y	Y	Web designer	47000	F	Y		
Sally	Sadik	F	12/03/1967	8	Y	N	Programmer	38000	F	Y		
Mustafa	Karwad	M	01/02/1984	4	N	Y	Systems analyst	54000	F	Y		
Alex	Gomaz	M	30/09/1993	0	N	N	Artist	41000	F	Y		
Bianca	Schastok	F	03/11/1980	0	N	N	Systems analyst	56500	F	Y		
Vyoma	Pathak	F	14/12/1956	1	N	Y	Technician	41000	F	N		
Nakul	Borade	M	22/06/1960	5	Y	Y	Web designer	45000	F	Y		
Rachel	Liu	F	13/12/1961	7	Y	Y	Programmer	38900	F	N		
Sho Ling	Wong	F	17/09/1978	9	Y	Y	Animator	28000	P	N		
Chloe	Burns	F	10/02/1972	5	N	Y	Security analyst	52000	F	Y		
Rachel	Hughes	F	16/09/1991	0	N	N	Network administrator	34200	P	N		
Grace	Hughes	F	25/12/1965	4	N	Y	Director	78000	F	Y		
Marzena	Jankowski	F	31/12/1955	2	N	Y	Admin clerk	32000	P	Y		
Bishen	Singh	M	16/01/1958	0	N	N	Assistant network manager	39500	F	N		
Raol	Ncube	M	12/01/1974	5	Y	Y	Director	87000	F	Y		
James	Murphy	M	30/06/1964	4	N	Y	Admin clerk	26000	P	N		
Rajan	Uppal	M	22/08/1977	5	Y	Y	Web designer	14000	P	Y		
Hamid	Zadeh	M	03/01/1992	10	Y	Y	Trainee analyst	23000	F	Y		
Kevin	Fortuni	M	30/09/1990	7	Y	N	Trainee analyst	26000	F	Y		
Maria	Fortuni	F	16/06/1989	11	Y	Y	Technician	27000	F	N		
Rupinder	Singh	M	29/05/1990	6	Y	N	Finance clerk	24900	P	N		
Emily	Wilson	F	27/12/1989	4	Y	N	Finance clerk	37000	F	Y		
Osama	Diad	M	03/11/1993	2	N	N	Trainee network engineer	23000	P	N		
Ahmed	Fathy	M	23/12/1990	7	Y	Y	Network engineer	41800	F	Y		
Hassan	Sheata	M	09/11/1989	6	Y	N	Network engineer	43000	F	Y		
Abdullah	Nordin	M	02/01/1969	2	N	N	Web designer	42000	F	N		
Fay	Hoy	F	09/10/1988	8	Y	Y	Receptionist	21300	P	N		
Robert	Marley	M	17/05/1965	0	N	N	Marketing administrator	21000	P	N		
Hassouneh	Al Sheikh	M	01/01/1990	5	Y	N	Marketing administrator	27600	F	Y		
Samantha	Jackson	F	09/12/1970	7	Y	Y	Programmer	30500	F	Y		
Mia	Hamm	M	06/11/1978	2	N	Y	Finance clerk	12000	P	N		

Page 1 of 1

6 Save the spreadsheet using the filename **Database of employees with header and footer**.

Presenting data

In this section you will learn about altering the appearance of the spreadsheet by making use of different fonts and font sizes, the use of styles such as bold, underline, etc., the use of colour, aligning data in cells, etc. You will be given precise instructions in the examination as to how to present the data in the spreadsheet.

Adjusting the display features in a spreadsheet

The ways of altering the appearance of text in a spreadsheet can be accessed using the Home tab. In order to alter the data in the spreadsheet you click on the cell or highlight the data to have its appearance changed if it runs over more than one cell and then click on one or more of the icons in the font section.

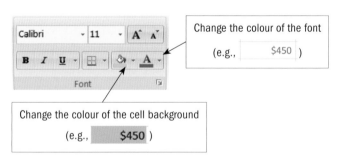

Change the colour of the font (e.g., $450)

Change the colour of the cell background (e.g., $450)

Always choose background and font colours carefully. Make sure there is enough contrast between them. If you print the spreadsheet out in black and white, there may not be enough contrast to be able to read the data.

Choosing a font (changing the shape of the letters and numbers (called fonts))

If you don't like the font that the computer has automatically chosen (called the default font) they you can change it. To do this you select the text you want to change and then click on the font box.

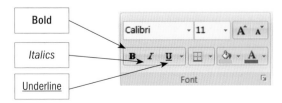

Bold

Italics

Underline

Choose a font from the drop down list of fonts when you click here.

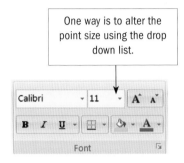

Altering the size of the font

There are a few ways to alter the size of a font.

One way is to alter the point size using the drop down list.

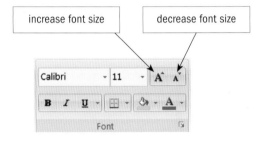

Another way is to use these two icons:

increase font size

decrease font size

Using the Format Cells menu

You can format cells using the Format Cells menu by right clicking and then selecting Format Cells. The selected cells in the spreadsheet can be formatted using this menu:

Adding colour/shading to enhance the spreadsheet

To add colour or apply shading to cells, you first have to select the cells by left clicking on them and dragging until the cells are highlighted. Then right click on the selected cells and the following menu appears, from which you need to select Format Cells:

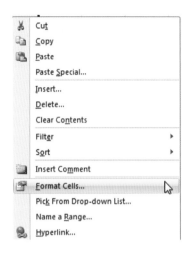

The Format Cells window appears:

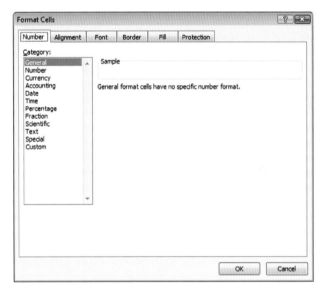

Select the Fill tab and the following window appears where you can select colours and patterns:

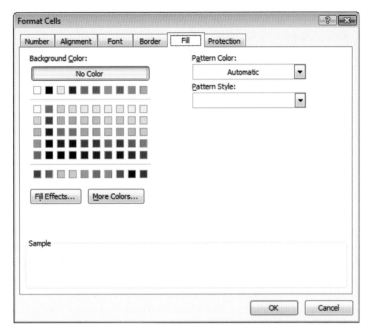

If you want to adjust the column widths automatically so they accommodate the width of the field names and the data in the fields, there is the following quick way: Click on the following symbol at the top left of the spreadsheet 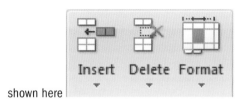. This will select the entire spreadsheet. When this is done, you will see all the spreadsheet cells change to light blue. Position the cursor between two of the columns (any will do) and double click the left mouse button. You will see the columns automatically change so they fit all the contents.

Hiding rows or columns in a spreadsheet

You may be asked in the examination to show a spreadsheet with rows and columns of data hidden.

1   First select the row(s) or column(s) you want to hide.

2   Click on **Home** then click on Format from Cells group

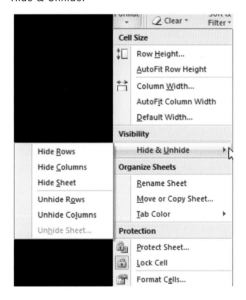

shown here

The following menu appears from which you should select Hide & Unhide.

In the menu on the left you can now decide whether you want to Hide columns or rows.

▶▶ Notice that the above menu also allows you to adjust the row height and column width.

3   To unhide you do not need to select any cells first. Simply go to the Format Cells and then click on either Unhide Rows or Unhide Columns.

Aligning data and labels

Data in cells is normally positioned according to the following:

▶▶ Numbers are aligned to the right in a cell.
▶▶ Text is aligned to the left in a cell.

To align cells click on the Home tab and then look at the alignment section of the screen.

To align the data or labels in a cell, select it and click one of these icons:

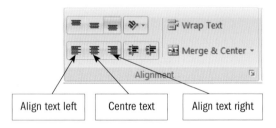

| Align text left | Centre text | Align text right |

Adjusting column width and row height

To adjust the width of a column position the cursor on the line between the column letters like this:

Keeping the left mouse button pressed down, drag the cursor to the left or right to change the width.

To adjust the row height, move the mouse onto the line joining the rows like this:

Keeping the left mouse button pressed down, drag the cursor up or down to change the row height.

Presenting results as graphs and charts

Tables of figures can be hard to understand so it is much easier to show these figures as a picture using graphs and charts. Using graphs/charts makes it easy to:

- ▸▸ Look at the relative proportions of items
- ▸▸ Spot any inconsistencies in the data
- ▸▸ Spot trends such as profits increasing or decreasing
- ▸▸ See the biggest or smallest reading.

There many graphs and charts you can produce using the spreadsheet software Excel, so it is important to pick the one that is most suitable.

Pie charts

These are good for displaying the proportion that each group is of the whole. For example, you could show a class's crisp preferences using a pie chart.

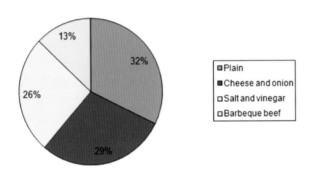

Group 7A's crisp preferences

Bar charts

Good for displaying the frequency of different categories. Here is a bar chart to investigate the types of vehicle using a certain road as a shortcut.

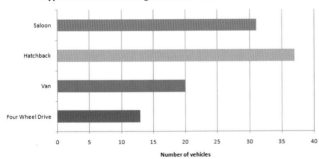

Types of vehicle taking a certain road as a shortcut

Bar charts in Excel can be:

- ▸▸ Vertical bar charts, or column charts as they are called in Excel, display the bars vertically.
- ▸▸ Horizontal bar charts are called simply bar charts in Excel and they show the bars in a horizontal direction across the screen.

Column charts

Column charts are used to compare values across different categories. For example, you could use a column chart to compare sales of cars for the first six months of the year.

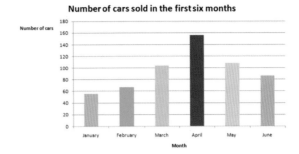

Number of cars sold in the first six months

Scattergraphs

To see how closely, if at all, one quantity depends on another. This is called correlation. For example, you might start with the hypothesis that someone who is tall will have bigger feet. You would then collect height and shoe size data and then plot the pairs of values.

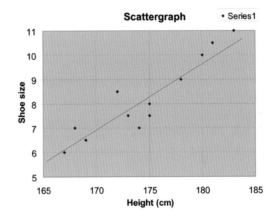

Line graphs

These can be used to show trends between two variables. The graph shown here shows how the value of a car falls over the first four years from new. Here the value is plotted against the time in years.

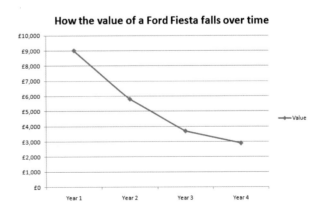

How the value of a Ford Fiesta falls over time

Activity 14.18

Producing a scattergraph

In this activity you will learn the following skills:

▶▶ Select data to produce a graph or chart

▶▶ Label the graph or chart with a title, legend, and axes

A driving school advertises for pupils and they would like to answer the following what if question: What if we spend more on advertising? Will we get more pupils?

They have collected the following data and put it into a table:

Amount spent on advertising per week ($)	Number of new pupils per week from adverts
20	1
30	2
40	4
50	4
60	5
70	6
80	6
120	9
150	10
160	12

This data is quite hard to interpret, so they have decided to present it as a scattergraph. This will enable them to see the relationship (if there is one) more clearly.

1 Firstly enter the data into the worksheet like this:

	A	B
	Amount spent on advertising per week ($)	**Number of new pupils per week from adverts**
1		
2	20	1
3	30	2
4	40	4
5	50	4
6	60	5
7	70	6
8	80	6
9	120	9
10	150	10
11	160	12
12		

Check that you have centred the data in both columns.

2 Select the data by clicking and dragging the mouse from cells A1 to B11. The selected area will be shaded.

3 Click on Insert in the toolbar and notice the charts section shown:

Click on 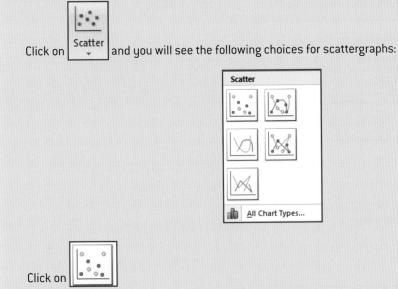 and you will see the following choices for scattergraphs:

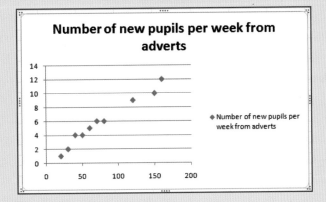

Click on

4 The scattergraph appears on the spreadsheet like this:

Number of new pupils per week from adverts

◆ Number of new pupils per week from adverts

5 Single click on the border of the chart to select it.

You will see │ Layout │ on the toolbar. Click on this.

A whole series of aspects of the layout now appear as shown here:

In the following section click on Axis Titles.

6 The following menu appears where you can choose which axis you want to add a title to:

Primary Horizontal Axis Title ▶
Primary Vertical Axis Title ▶

Click on Primary Horizontal Axis Title. The menu shown here appears:

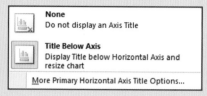

None
Do not display an Axis Title

Title Below Axis
Display Title below Horizontal Axis and resize chart

More Primary Horizontal Axis Title Options...

Click on Title Below Axis.

7

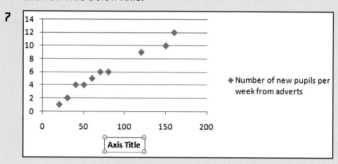

You will now see the Axis Title box appear. Change the text in this box to read "Amount in Dollars spent on advertising each week".

Your scattergraph should now look like this:

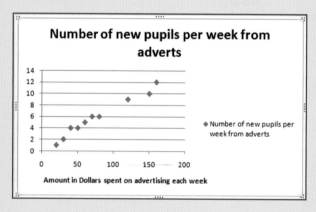

8 Put a title "Number of new pupils per week" on the vertical axis. From the list of options, choose to put the axis title horizontally next to the axis. Remember you will need to select the chart before you will see the layout tab on the toolbar.

Your chart should now look like this:

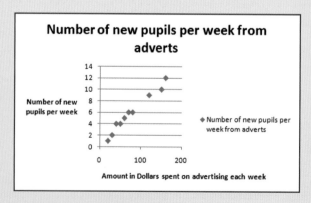

9 Save your spreadsheet using the filename **Scattergraph**.

Using contiguous and non-contiguous data

Contiguous data is data in columns or rows that are next to each other. Non-contiguous data means data in rows or columns that are not next to each other. For the examination, you will have to produce graphs and charts using both types of data.

Activity 14.19

Producing a column chart

In this activity you will learn the following skills:

▸▸ Copy and paste data
▸▸ Widen columns
▸▸ Change row height
▸▸ Use contiguous and non-contiguous data to produce charts
▸▸ Label the chart with title, legend, category, and value axes
▸▸ Scale and preview a printout

The following table has been created using word-processing software in a document about nutrition and healthy eating. The writer of the article would like to produce a bar chart that can be used to compare the protein and fat in each of the different foods.

Product	Oven chips	Soup	Yoghurt	Salad Cream	Corn Flakes
Amount (g) of protein per 100g	2	1	5	1	9
Amount (g) of fat per 100g	5	3	1	10	6

1 Load the word-processing software Word and open the file called **Table showing fat and protein**.

Check your table looks the same as that shown above.

2 Rather than type the data in again you have to select all the data in the table by clicking on the text keeping your finger on the left mouse button, dragging until all the data is highlighted like this:

Product	Oven chips	Soup	Yoghurt	Salad Cream	Corn Flakes
Amount (g) of protein per 100g	2	1	5	1	9
Amount (g) of fat per 100g	5	3	1	10	6

With the cursor positioned somewhere on the table, right click the mouse and in the menu that appears select **Copy**. This copies the selected data to the Clipboard.

3 Load the spreadsheet software Excel.

When the spreadsheet grid appears, click on cell A2 as this is where we want the data we are copying to start.

Right click the mouse and from the menu that appears select **Paste**.

Check the table has appeared like this:

	A	B	C	D	E	F
1						
2	Product	Oven chips	Soup	Yoghurt	Salad Cream	Corn Flakes
3	Amount (g) of protein per 100g	2	1	5	1	9
4	Amount (g) of fat per 100g	5	3	1	10	6

4 Adjust the column width by clicking and dragging the line between the columns like this:

	A	B
1		

to the right.

The final spreadsheet should look similar to this:

	A	B	C	D	E	F	G
1							
2	Product	Oven chips	Soup	Yoghurt	Salad Cream	Corn Flakes	
3	Amount (g) of protein per 100g	2	1	5	1	9	
4	Amount (g) of fat per 100g	5	3	1	10	6	
5							

5 Adjust the row height by right clicking and dragging the lines between the rows like this

3	Amount (g) of protein per 10

upwards.

The spreadsheet should look similar to this:

	A	B	C	D	E	F	G
1							
2	Product	Oven chips	Soup	Yoghurt	Salad Cream	Corn Flakes	
3	Amount (g) of protein per 100g	2	1	5	1	9	
4	Amount (g) of fat per 100g	5	3	1	10	6	
5							

6 You are now going to draw a vertical bar chart (called a column chart) using some of the data in the spreadsheet (i.e., the first two rows). Because these two rows are next to each other, the data in them is called **contiguous data**.

Select the data by left clicking on cell A2 and then dragging to the end of the data we want to use in cell F3. This area will now appear highlighted.

7 Click on the | Insert | tab in the toolbar and then select | Column |.

Choose the first chart in the 2D column section.

The column chart is now drawn:

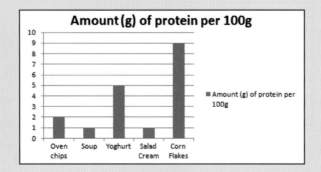

Notice the following about this vertical bar chart (or column chart):

The **category axis** (i.e., the horizontal axis) has the categories (i.e., the names of the items) on it.

The vertical axis, called the **value axis**, has the numbers relating to the category on it.

The **legend** is the explanation (here on the right of the chart) that explains what the height of the bars represents.

The **title** of the chart gives a few words to explain what the chart shows (i.e., Amount (g) of protein per 100g). This is only a suggested title and you can change this.

8 You are now going to change the title of the chart.

Click on the title to select it. When the item is selected it will appear like this:

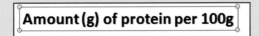

Replace the text in title with the text shown below.

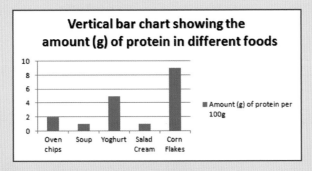

9 Rather than use a legend you are going to label the value axis.

Select the legend by clicking on it and then click on 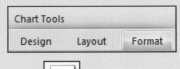 and press the backspace button on your keyboard to delete it.

10 You are now going to add a title to the value axis.

Click on the edge of the chart to select the entire chart.

The chart tools will appear like this:

Chart Tools

Design   Layout   Format

Click on Layout Layout and then on Axis Titles ▾ and then on "Choose Primary Vertical Axis Title" and finally choose "Horizontal Title".

Notice the text box next to the vertical (value) axis.

Delete the text and replace it with the text "Grams of protein".

Your chart should now look like that shown below:

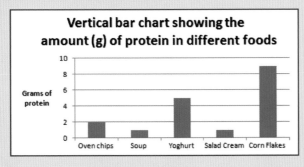

11 Save your spreadsheet using the filename **Food bar chart version 1**

12 You are now going to draw a similar vertical bar chart, this time showing the amount (g) of fat in different foods.

Look at the data in the spreadsheet:

1						
2	Product	Oven chips	Soup	Yoghurt	Salad Cream	Corn Flakes
3	Amount (g) of protein per 100g	2	1	5	1	9
4	Amount (g) of fat per 100g	5	3	1	10	6

You need to use the first row of data and the third row of data. This data is **non-contiguous** because the two rows are not next to each other.

You need to select the cells from A2 to F2 by clicking and dragging. Press the Ctrl key down and keep it pressed down click on cell A4 and drag the mouse as far as cell F4. Row 2 and row 4 should now be highlighted like this:

	A	B	C	D	E	F
1						
2	Product	Oven chips	Soup	Yoghurt	Salad Cream	Corn Flakes
3	Amount (g) of protein per 100g	2	1	5	1	9
4	Amount (g) of fat per 100g	5	3	1	10	6

13 You now have to produce the vertical bar chart in a similar way to the way you did the one for the amount of protein. Make sure you alter the text to that shown below.

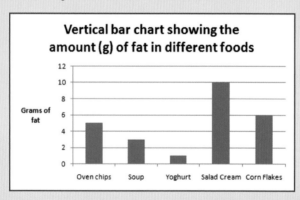

You will probably need to move the charts so they appear next to each other. To do this click on the border of the chart; when a four headed arrow appears, hold the right mouse button down and drag the chart into the correct position.

Here is what your final charts may look like:

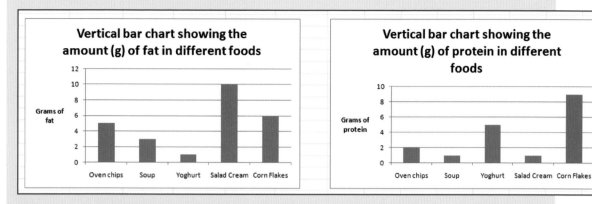

14 Save your spreadsheet using the filename **Food bar chart version 2**

15 You are now going to print the spreadsheet (table of data and the two charts).

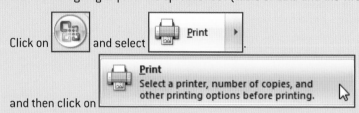

Click on [ ] and select [Print] ▸ .

and then click on

Print
Select a printer, number of copies, and other printing options before printing.

16 The following Print window opens:

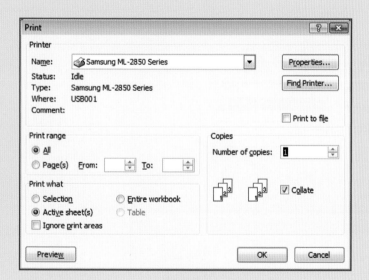

Click on [Properties...] when the following properties window appears:

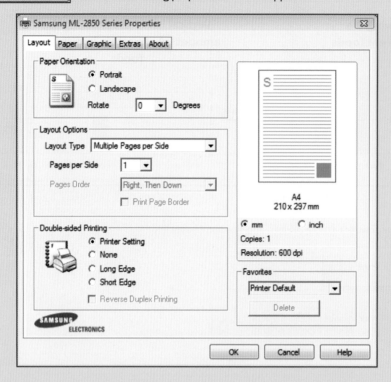

As this spreadsheet is wider than it is tall, it would be better to print it using landscape paper orientation. Click the radio button for Landscape.

Click on the Paper tab and in the Scaling Printing section select **Fit to page**.

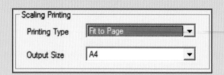

Click on OK and you will return to the Layout information. Click on OK again and you are taken to the main Print menu shown here:

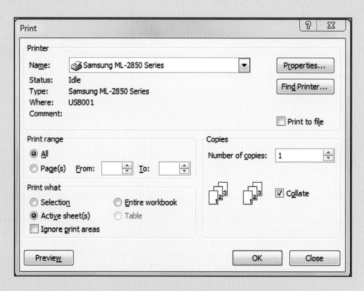

Click on Preview to check that the printout is on one page in landscape orientation.

Click on the Print icon and a printout is produced.

Activity 14.20

Producing a pie chart

In this activity you will learn the following skills:

▶▶ Create a pie chart
▶▶ Label pie chart segments with percentages
▶▶ Add a title

The table below shows the type of heating used in 100 houses.

Type of heating	Number of houses
Solar	16
Wood	24
Coal	12
Gas	8
Electricity	10
Oil	30

1 Load the spreadsheet Excel and key in the data shown in the above table.

Verify your data by proof-reading (i.e., checking your data is exactly the same as that shown in the table).

2 Select all the data in the table by clicking and dragging from cells A1 to B7.

3 Click on 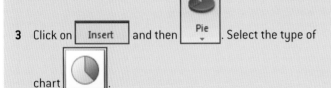 Insert and then Pie . Select the type of chart .

The pie chart appears like this:

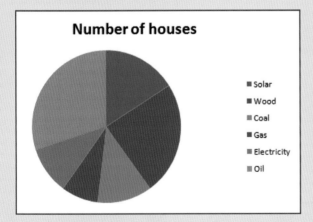

4 There are a number of problems with the chart as it is shown here:

▶▶ The title does not correctly express the purpose of the chart.
▶▶ It could do with more explanation of the legend.

Click on the chart title and change it to "**Pie chart showing the type of heating used in a sample of 100 houses**".

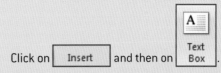

Click on Insert and then on Text Box .

Click and drag to create a text box in a position just above the legend as shown below.

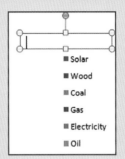

Enter the text "Type of heating" in the text box.

Centre the text in the text box by highlighting the text and clicking on ≡ .

Your chart should now look like this:

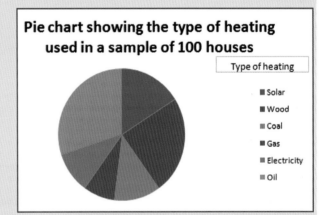

5 You are now going to put labels on the segments (these are the slices of the pie). You can add the values themselves (i.e., the actual number that represents a segment) or you can add percentages.

Click on the chart area to select it. Now click on

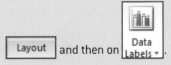

Layout and then on Data Labels .

On the pull-down menu that appears, click on

More Data Label Options... .

The following window appears:

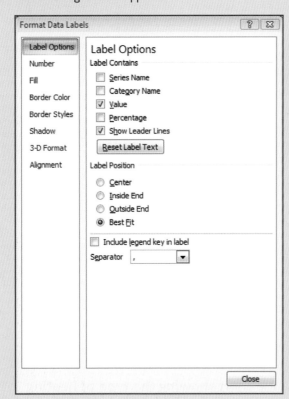

Notice how the actual values are now displayed on the segments.

Click on the tick in the Value box to deselect it and click on Percentage.

Notice that the percentage values have been added to the segments:

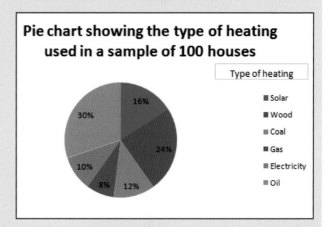

6 Save your spreadsheet using the filename "**Heating pie chart**".

Changing the appearance of the pie chart

In this activity you will learn the following skills:

▶▶ Change the colour scheme or patterns of a chart

▶▶ Extract a pie chart segment

1 Load the spreadsheet software and open the file created in the last activity called "**Heating pie chart**" if not already opened.

2 Click on the border of the chart to select it.

You are now going to change the colour scheme for the chart.

Click on Design in the Chart tools section.

Notice the following examples of chart colour schemes:

If you are going to print the pie chart in black and white, the first colour scheme should be used (i.e., shades of grey) as this will give enough contrast between the segments.

Choose the pastel green colour scheme by clicking on it.

3 You can also add patterns to a chart.

Right click on a segment inside the pie chart and the following box appears:

Select the last item Format Data Series… and the following window appears:

Choose Fill and then click on Picture or texture fill:

Click on the drop down arrow for Texture

Texture: [icon ▼] and the following textures appear.

Choose a suitable pastel colour such as "Blue tissue"

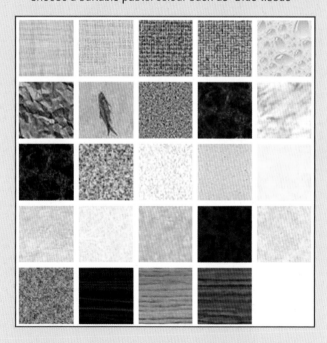

You are now going to emphasize a segment by moving it out away from the rest of the pie chart.

Double left click on the segment with the value 24% to select it. When this segment has been selected you will see the small circles appear like this:

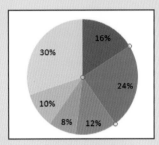

Right click and a menu appears from which you need to select Format Data Point:

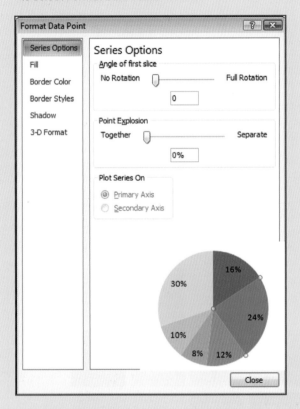

Move the slider on the Point Explosion until the number reads 25% like this:

Now experiment by changing the Angle of first slice and the Point Explosion.

You can then save the spreadsheet using the filename **Exploded pie chart**.

Activity 14.22

Producing line graphs

In this activity you will learn the following skills:

▸▸ Produce a line graph
▸▸ Add a second axis
▸▸ Change the axis scale maximum and minimum

Line graphs are useful to show trends. For example, you can show the trends in hours of sunshine per day and the maximum daily temperature for each month over a year.

You are going to produce two such line graphs for Dubai, UAE.

1 Load the spreadsheet software Microsoft Excel and open the file **Dubai weather table**.

   Check you have the following data shown on your screen:

	A	B	C
1			
2	Month	Average hours of sunshine per day	Average max daily temp °C
3	Jan	8	23
4	Feb	9	24
5	Mar	8	27
6	Apr	10	30
7	May	11	34
8	Jun	11	36
9	Jul	11	38
10	Aug	10	39
11	Sep	10	37
12	Oct	10	33
13	Nov	10	31
14	Dec	8	26

2 Select the data to be used to create the graph by clicking on cell A2 and dragging to cell C14 and then

   click on Insert and finally on Line which tells the computer to create line graphs.

3 Select the first line graph from the menu by clicking on the icon:

4   The pair of line graphs is drawn on the same axes like this:

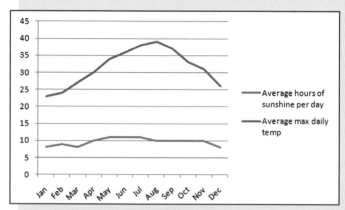

5   The problem with putting two graphs on the same set of axes is one can appear a bit flat like the one here for the average hours of sunshine per day. It is possible to have two vertical axes with each axis referring to its own line graph.

Right click on the blue line, as this is the line we want to create the secondary axis for. From the menu that appears, select Format Data Series.

6   The following box appears where you need to select Secondary Axis. This tells the computer that you want to create a secondary axis for the data represented by the blue line that has been selected:

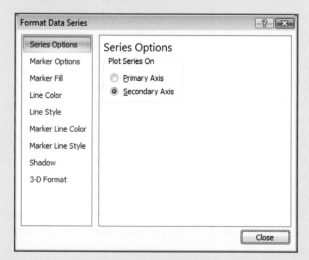

Click on Close.

7   The secondary axis now appears like this:

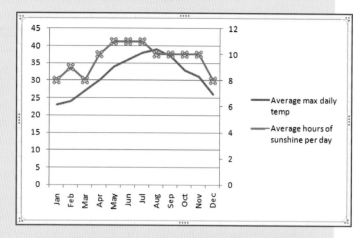

There is still a slight problem. As you can see Excel scales the axes automatically but it is sometimes better to adjust this yourself so that the graphs are stretched out a bit more.

You are now going to scale the axes yourself.

Right click on the secondary axis (i.e., the one on the right) and from the menu that appears select Format Axis and the following box appears:

If you look at the data for the hours of sunshine either in the original table or on the graph you will see that the largest data item is 11 and the smallest is 8.

In the Axis Options section:

In the Minimum section click on Fixed to select it and change the value to 7.

In the Maximum section click on Fixed and change the value to 12.
Click on Close and the changes are made.

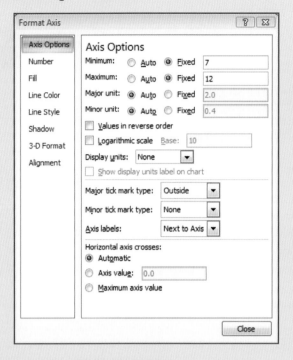

8 The scale on the primary axis needs adjusting slightly.

Right click on the primary axis (i.e., the one on the left) and from the menu that appears select Format Axis and you then need to change the settings to the following:

Axis Options

Minimum:	Auto	◉ Fixed	15.0
Maximum:	Auto	◉ Fixed	40.0

Your chart should now look like this:

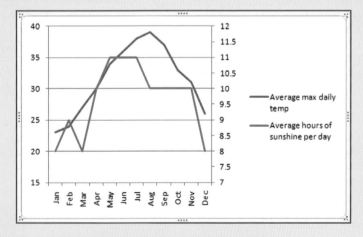

9 You now need to add a title for the entire chart and a title for each axis.

Before you do this, make sure that you click on the border of the chart to select it.

To add the chart title: Click on Layout on the toolbar and then on Chart Title ▾.

Choose "Above Chart" from the list of options. A textbox appears above the chart into which you should enter the text **Monthly weather data for Dubai, UAE**.

Now add the axis titles as follows:

To add the chart title: Click on Layout on the toolbar and then on Axis Titles. Select Primary Vertical Axis Title and then select Rotated Title.

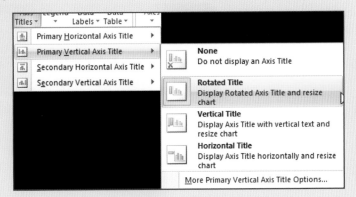

Enter the text **Temperature** in the text box next to the axis.

To add the chart title: Click on Layout on the toolbar and then on Axis Titles. Select Secondary Vertical Axis Title and then select Rotated Title.

Enter the text **Hours of sunshine**

You will now need to increase the size of the chart by clicking on the corner of the border and dragging the two-headed arrow that appears until the graph is about 1.5 times its original size.

Your chart should now look like this:

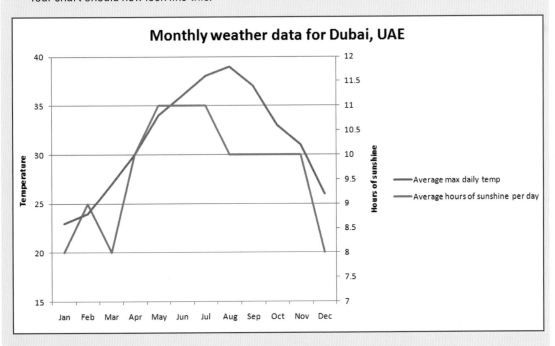

10 Save your chart using the filename **Chart showing monthly weather data for Dubai**.

11 Print a copy of the spreadsheet out in suitable orientation and on a single sheet of paper.

15 Web site authoring

In this chapter you will be learning how to create a structured web site with stylesheets, tables and hyperlinks. You will be using a special code for this called HTML. There are other ways to create web pages using special web-authoring software such as Adobe Dreamweaver or Microsoft FrontPage. In this book you will only be using a text editor to create the HTML code and then use web browser software to view the web page you have produced.

The key concepts covered in this chapter are:
▸▸ Use stylesheets
▸▸ Create web pages
▸▸ Use images on web pages

HTML (HyperText Markup Language)

HTML is short for HyperText Markup Language and is the special code that is used for making web pages. HTML consists of special markers called tags that tell the computer what to do with the text, images or links that are entered. It could tell the computer to present the text in a certain way, or size and align an image on the page. For example, the tags could tell the computer that the text being entered is intended to be a heading or to make a certain block of text bold.

HTML is a text file, just like MS Word, except that it contains special markers called tags. HTML can be entered into a text editor, which enables tags to be entered as well as content to which the tags apply. You will see how to use a simple text editor for entering and editing HTML later.

Important note: HTML is not a programming language as such. It just tells the computer how to display text and pictures on web pages.

⊙ **KEY WORDS**

HyperText Markup Language (HTML) used to create documents on the World Wide Web. You use it to specify the structure and layout of a web document.
Tags special markers used in HTML to tell the computer what to do with the text. A tag is needed at the start and end of the block of text to which the tag applies.

Browser and editor software

Browser software is responsible for requesting text and graphics on web pages stored on servers and then assembling them for display. In this chapter, web browser software will be used to view the web pages you create. Editor software is used for creating and amending text. In this chapter, you will be using a text editor for the preparation of HTML web pages.

Here are the steps involved in creating and viewing web pages created using HTML:

1 Load the text editor software.
2 Type in the HTML code.
3 Save the HTML file.
4 Load the browser software.
5 View the web page and see if there are any mistakes in the code.
6 Go back to the HTML code using the editor software.
7 Edit the HTML code using the editor.
8 Save the HTML file.
9 View the web page using the browser.
10 If there are any more corrections needed go to step 6.
11 Save the final HTML code.

Setting up a relevant folder structure

Before you start work on this chapter it is important that that you create a folder structure to hold all the files you will be making in the activities.

To create a folder structure for your work follow these steps:

1 Click on  at the bottom left of your screen.

2 Select Computer from the menu

on the right

3 You will be presented with a screen like this showing the hard disk drives and devices with removable storage.

Click on the device where you want the HTML files to be stored. In most cases this will be one of the hard drives or a flash drive. Your teacher will tell you on which drive/device the files are to be stored.

4 Double click on the name of the device and you will be presented with a list of folders currently on that device.

5 Right click on the white area around the file list and the following menu appears:

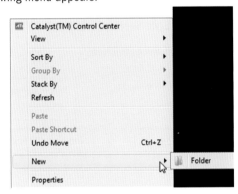

Click on New followed by Folder.

6 A box appears like this:

Delete the text inside the box and replace it with the name HTML like this:

7 The folder now appears in the list of folders like this:

IGCSE
Review and marketing
HTML
Ro

Click on the folder HTML.

8 The empty folder appears like this:

9 Click on  and then select New Folder from the menu.

The sub-folder appears like this as a folder within the folder called HTML. Note that it is called a sub-folder because it is a folder within a folder and lower down in the hierarchy.

Click once on the New Folder.

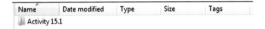

You can now type in the name for this sub-folder.

In this case we will call this sub-folder Activity15.1

Type in Activity 15.1 and the sub-folder appears like this:

Name	Date modified	Type	Size	Tags
Activity 15.1				

You can now add sub-folders for any of the other Activities as they are needed.

Saving all the web files in the same folder or sub-folder
When a web page is created, all the files used in the web-page should be stored in the same folder or sub-folder as the HTML file. This means that if a web page contains an image then the image file should be stored in the same folder or sub-folder as the HTML file.

The structure of an HTML document
HTML documents need to be structured in a certain way. Here are the basics of the structure:

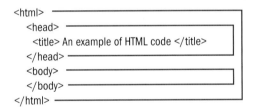

Look at the above section of code carefully. You will see the way the tags are nested within each other (i.e., shown by the blue lines) and also how indentation is used to make it easier to see the various parts of the code.

The tags are the words enclosed between the < and > signs. They are HTML instructions and they tell the computer how to display or format the text.

Here is what each tag in this section of code does:

▸▸ <html> and </html> tells the computer that we are creating an HTML document. The <html> tag tells the computer that the document is starting and the </html> tells the computer that the document has ended.

- <head> and </head> is a section where you provide information about your document.
- <title> and </title> tells the computer that you want to put the text between the tags as a title. As the title is part of the "head" section it needs to be inserted after the opening head tag <head> and before the closing tag </head>. The nesting of tags in this way is very important as not nesting starting and closing tags is a frequent cause of problems when writing HTML. The text used as the title does not appear on the web page but instead is shown in the browser's window. If a user of the web pages adds the page to their favourites, this title text is used as the name of the page.
- <body> and </body> is a section where you insert the content of your document and visual information about how it is to appear on the web page.

More code and what it does
- <h1> and </h1> tells that you want the text to be a heading of type h1. There are other headings h2, h3, etc., to choose from. h1 headings are larger in size compared to h2 headings and so on. The text you want on the page is placed between these two tags like this: <h1>This text is size h1</h1>
- <p> and </p> mean start and end a new paragraph. A break between the paragraphs is inserted that is approximately two lines in length. The text for the paragraph is placed between the two tags.

More tags and what they do
Here are some other tags and what they do:

- <b> and </b> means start bold and stop bold style text.
- <u> and </u> means start underline and stop underline style text.
- <i> and </i> means start italics and stop italics style text.
-
 means insert a line break. There is no closing tag for this. This creates a smaller gap than the <p> tag.
- <hr> means insert a horizontal line and there is no closing tag.

Aligning text using HTML
Text can be aligned on the page to the right, left or centre. The text has to be in a separate paragraph in order for it to have specific alignment attributes attached to it. Here is an example of HTML using different alignments:

```
<html>
    <body>
        <p align="left">This is an example of left aligned text.</p>
        <p align="center">This text has been centred.</p>
        <p align="right">This text has been right aligned.</p>
    </body>
</html>
```

Creating an external stylesheet

Before the introduction of stylesheets, if you had a web site containing 50 pages and wanted to change the text for all the h1 headings from black to blue, you would need to look for all the h1 tags on all the web pages and alter them manually. This would be time consuming.

Luckily we now have stylesheets and you would only need to make the change once in the stylesheet and all the text for the h1 tags would change from black to blue automatically.

It is important to note that stylesheets are not HTML documents and do not use tags and you will never see < and > used in stylesheets. Also the files used to save stylesheets do not have the file extension .htm but instead use .css

The use of stylesheets has the following advantages:

- They save time – you only have to change things in one document (i.e., the stylesheet).
- They help give a web site a consistent look (e.g., heading sizes, fonts, font sizes, etc., will be consistent across all pages).

There are a few rules you need to obey when creating a stylesheet using CSS (Cascading Style Sheets).

CSS codes have the following structure:

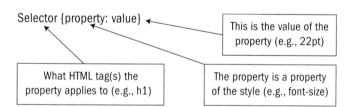

Selector {property: value}

This is the value of the property (e.g., 22pt)

What HTML tag(s) the property applies to (e.g., h1)

The property is a property of the style (e.g., font-size)

So to change the size of the font used for the h1 heading we would use the following CSS stylesheet:

h1 {font-size: 22pt}

To understand stylesheets you need to look at a few examples:

Here two different styles are applied to the h1 heading:

h1 {text-align: center, color: blue}

Notice the American spelling of both "Center" and "Color" which you must use.

This stylesheet sets the heading h1 to the colour blue and also aligns the text centrally. Notice the curly brackets at the start and end of the stylesheet and also notice that there is a comma separating the properties.

If you had a web site and used a stylesheet and wanted all occurrences of the h1 heading to be changed from blue to red, you would only need to change the stylesheet slightly to this:

h1 {text-align: center, color: red}

Activity 15.1

Creating a simple web page using HTML

In this activity you will create a very simple web page using HTML and you will learn the following skills:

▸▸ How to use the editor to create HTML code

▸▸ How to use simple tags to create a web page

▸▸ How to alter the sizes of headings using h1, h2 and h3

▸▸ How to save an HTML document

1 From your operating system screen, click on the Start

button  then click on All Programs and then Accessories and then on Notepad.

2 Notepad is an editor and we can use it to put the HTML together.

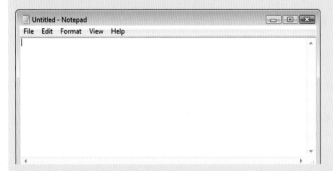

Type the following text into the editor exactly as it is shown here:

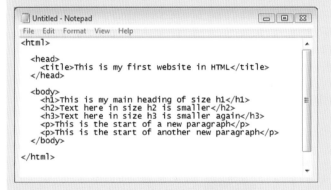

3 We need to save this file as an HTM file. To do this click on File and then Save As.

A window like this appears.

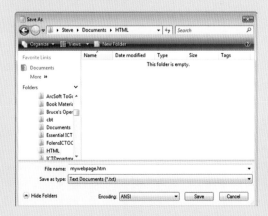

You need to find the folder created for saving your HTML documents. Once you have done this, enter the filename **Mywebpage.htm** and then click on Save.

Important note: You must always remember to add the .htm file extension to the filename, as if you do not do this then the browser you use to open the web page will not recognize it as a web page.

Check that your file has been saved correctly like this:

Name
🅔 Mywebpage

Notice the browser symbol is shown next to the filename. If you do not see this symbol then you have probably missed out the .htm file extension and you will have saved the file as a text file by mistake.

4 To open the HTML file you have just saved it is necessary to use a web browser program.

Load the web browser program you usually use (e.g., Internet Explorer).

5 Click on File and then Open.

You will now need to find the HTML folder.

Once you have found the folder click on the file Mywebpage.

Your web page will now load and you can see the results:

This is my main heading of size h1

Text here in size h2 is smaller

Text here in size h3 is smaller again

This is the start of a new paragraph

This is the start of another new paragraph

You can now close this web page.

There are two steps to create an external stylesheet:

1 Create the external stylesheet in .css format.
2 Link the stylesheet to the HTML pages you want it to apply to.

Font families

There is a problem with some fonts because not all web browsers can display them. This means when specifying fonts we need to produce a prioritized list of fonts. This means we do not just give one font and instead give the choice of two or more. What then happens is that the web browser looks along the list of fonts starting from the first one that is specified and if it cannot be used it then moves onto the next font in the list and so on.

There are two ways of specifying a font:

▶▶ By the name of the font such as Arial, Times New Roman, Calibri, etc.
▶▶ By whether the font is serif or sans serif.

This stylesheet specifies the font families for the h1 headings:

h1 {font-family: Calibri, Arial, sans-serif)

This means that if Calibri can be used by the browser, it will be used and if not, Arial will be used and if this is not available then any sans serif font will be used.

It is always best to end a font family with either sans-serif or serif as these will always be available no matter which web browser is used to display the web page.

If you use the name of a multi-word font such as Times New Roman, the name of the font must be placed inside inverted commas like this:

body {font-family: "Times New Roman", Century, serif}

Here is a list of properties and the values they can have:

Property	Value
Font-style	normal
Font-style	italic
Font-weight	normal
Font-weight	bold
Font-size	12pt, 22pt, etc.
Text-align	left
Text-align	right
Text-align	center
Text-align	justify

Background colour

You can change the background colour in the stylesheet like this:

body {background: red}

or by using a code for red (#FF0000)

body {background: #FF0000}

Note the background colour of the body is being defined here.

There are a number of colour names you can use shown here:

black (#000000)	silver (#C0C0C0)	gray (#808080)	white (#FFFFFF)
maroon (#800000)	red (#FF0000)	purple (#800080)	fuchsia (#FF00FF)
green (#008000)	lime (#00FF00)	olive (#808000)	yellow (#FFFF00)
navy (#000080)	blue (#0000FF)	teal (#008080)	aqua (#00FFFF)

Note: the American spelling of the colour "Gray".

If you want to use a colour that is not one of the sixteen named colours, you can refer to a colour palette and then use the hexadecimal code for the colour.

Palette of colours

To get a palette of colours with their codes use the following web site:

http://www.webmonkey.com/2010/02/color_charts/

Activity 15.2

Creating a stylesheet

In this activity you will learn the following skills:

▶▶ Create a stylesheet in .css file format
▶▶ Link a web page to a stylesheet

In this activity you will be creating a stylesheet which will then be used by a section of HTML code to set the styles for a web page.

For this activity you will be creating two files:

A stylesheet using the .css file format

A web page using the .htm file format

Both of these files can be created and saved using a text editor such as Notepad.

1 From your operating system screen, click on the Start button 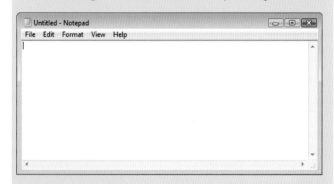 then click on All Programs and then Accessories and then on Notepad.

2 Notepad is an editor and we can use it to assemble the formatting instructions that make up the stylesheet.

Type the following text into the editor exactly as it is shown here:

body {background: #FEF76E; font-family: Arial, Verdana, sans serif }

h1 {font-family: Arial, Verdana, sans serif; font-size: 32pt; font-color: navy; font-weight: bold}

Important note: you must spell "colour" as "color" in both stylesheets and HTML code.

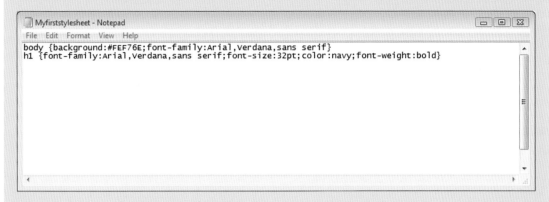

The text file needs to be saved in css format.

To do this click on File and then Save As

Call the file **Myfirststylesheet.css**

Click on Save.

3 With the stylesheet still being shown in Notepad, click on File and then New and type in the following HTML exactly as it is shown here:

<html>

    <head>

      <title>My first stylesheet</title>

      <link rel="stylesheet" type ="text/css" href="Myfirststylesheet.css">

    </head>

    <body>

      <h1>This heading should be point size 32 and colour navy and in bold</h1>

      <p>This is my first stylesheet so I hope it works</p>

    </body>

</html>

Notice the following step which is used to inform the computer that this page is to be linked to a stylesheet. Notice that the filename of the stylesheet is included so that the computer can find the stylesheet that is to be used with this web page.

    <link rel="stylesheet" type="text/css" href="Myfirststylesheet.css">

There are two important points about this line of HTML:

▸ It must be placed in the Head section of the HTML.
▸ The only part of the line you will need to change when you create your own HTML with stylesheets is the last part shown in blue here. All you need to change is the name of the file you are using for the cascading stylesheet.

Click on File and then Save As.

Check that you are saving the file in the folder you are using to hold your HTML and CSS files.

Type the following name for the file **Web page with a stylesheet.htm**

Click on Save.

You have now completed the HTML code of the web page including the link to the stylesheet.

4 Load your web browser software and click on File and then Open.

You will now need to locate the folder where your HTML and CSS files are stored.

You will see the HTML file for the web page displayed like this:

Click on this to open it.

The web page now opens in your browser like this:

This heading should be point size 32 and colour navy and in bold

This is my first style sheet so I hope it works.

You have now created a stylesheet and attached it successfully to the HTML code for a web page.

Tags to create lists

There are tags to create bulleted lists called unordered lists and numbered lists, which are called ordered lists.

The HTML to create a bulleted (unordered) list is as follows:

```
<ul>
<li> Saudi Arabia
<li> UAE
<li> Egypt
<li> Kuwait
<li> Argentina
</ul>
```

- Saudi Arabia
- UAE
- Egypt
- Kuwait
- Argentina

The HTML to create a numbered (ordered) list is as follows:

```
<ol>
<li> Saudi Arabia
<li> UAE
<li> Egypt
<li> Kuwait
<li> Argentina
</ol>
```

1 Saudi Arabia
2 UAE
3 Egypt
4 Kuwait
5 Argentina

Activity 15.3

Creating a stylesheet according to instructions

In this activity you will learn the following skills:

➤➤ How to change the appearance of a bullet in a list

➤➤ Use a stylesheet with common styles such as h1, h2, h3, p, and li

In the examination you may be asked to create a stylesheet for use within all the pages of a web site to aid consistency.

The stylesheet for a particular web site must contain the following styles:

h1 – red, sans-serif font, 34 point, centre-aligned, bold

h2 – dark blue, serif font, 20 point, left-aligned

h3 – green, sans-serif font, 12 point, left-aligned

li – blue, sans-serif, 10 point, bullet points, left-aligned

p – black, serif font, 10 point, left-aligned

1 Load Notepad and key in the following text to create the stylesheet shown here.

Tip: Once you have completed one line of the stylesheet, copy it and paste it for the second line. You then only have to make minor changes rather than type in the whole line, which will save time.

```
Style_sheet - Notepad

File  Edit  Format  View  Help
h1{font-family:Arial,sans-serif;color:red;font-size:34pt;text-align:center;font-weight:bold}
h2{font-family:"Times New Roman",Serif;color:navy;font-size:20pt;text-align:left}
h3{font-family:Arial,sans-Serif;color:green;font-size:12pt;text-align:left}
li{font-family:Arial,Sans serif;color:blue;font-size:10pt;text-align:left;list-style-type:circle}
p{font-family:"Times New Roman",Serif;color:black;font-size:10pt;text-align:left}
```

Look carefully at each line of the stylesheet and notice how the lines match the required styles.

Notice the way the font for each style is specified. First a font name is given (e.g., Arial or Times New Roman). If the font name is more than one word, the font name must be enclosed between quote marks like this "Times New Roman". After the font name and separated by a comma, the words either sans-serif or serif are used. This is so that if the required font is not available with the browser, then an alternative serif or sans-serif font can be used.

Notice that the colours of the fonts are specified like this:

color: navy

There are 16 colours that have names and these are; black, silver, gray, white, maroon, red, purple, fushia, green, lime, olive, yellow, navy, blue, teal, and aqua.

There are many other colours that you can use, but they are accessed using hexadecimal codes and further information about the colours and their codes can be found at: http://www.w3schools.com/html/html_colors.asp

In the style for the list, notice that the type of bullet used for the list can be specified like this:

list-style-type: circle

This means a circle is used for the bullet. Here is a summary of the types of bullet you can have:

List-style-type	Bullet appearance	Example
disc	Solid circle	• UAE • Egypt
circle	Circle	○ UAE ○ Egypt
square	Square	▪ UAE ▪ Egypt
decimal	Decimal number	1. UAE 2. Egypt

2 Save this stylesheet in your HTML folder using the filename **Style_sheet.css**

➡

3 You are now going to create the web page that makes use of the stylesheet you have just created.

At the top of the Notepad click on File and then New.

Type in the following HTML:

```
<html>
    <head>
    <title>Task setting up a stylesheet</title>
    <link rel="stylesheet' type="text/css" href="style_sheet.css"/>
    </head>
    <body>
        <h1>This is the h1 heading</h1>
        <h2>This is the h2 heading</h2>
        <h3>This is h3 heading</h3>
        <p>This is the text used for the majority of paragraphs on each page</p>
        <ul>
        <li>This is point one in the list
        <li>This is point two in the list
        <li>This is point three in the list
        <li>This is point four in the list
        </ul>
    </body>
</html>
```

Notice the fourth line where the stylesheet is linked to this web page.

Click on File and then Save As... and call the filename **Second web page with a stylesheet.htm**

4 Load your web browser software and click on File and then Open and locate your HTML folder and open the file **Second web page with a stylesheet.htm**

The web page now opens in your browser like this:

Task setting up a stylesheet

This is the h1 heading

This is the h2 heading

This is the h3 heading

This is the text used for majority of paragraphs on each page

- o This is point one in the list
- o This is point two in the list
- o This is point three in the list
- o This is point four in the list

Activity 15.4

Practising creating a stylesheet
In this activity you will practise the following skills:

▸▸ Create a stylesheet

In this activity you are required to create a stylesheet according to instructions. You may find it quicker to amend the stylesheet created in the last activity and then save the stylesheet using a different filename. You will then use the HTML file created in the previous activity to link to this new file. You will have to amend the HTML slightly as the filename for the stylesheet will have changed.

1 Create a stylesheet for use by a web page. The styles for this stylesheet are as follows:

   h1 – black, serif font, 28 point, centre-aligned, bold

   h2 – red, serif font, 20 point, centre-aligned, italic

   h3 – green, sans-serif font, 12 point, right-aligned

   li – black, sans-serif, 10 point, square bullet points, left-aligned

   p – blue, serif font, 10 point, left-aligned

Save this stylesheet using the filename **Style_sheet_act4.css**

2 Load the web page created in the last activity called filename **Second web page with a stylesheet.htm** into the Notepad. With the HTML shown in the Notepad change one of the instructions so that web page now uses the stylesheet with filename **Style_sheet_act4.css**

Save this HTML using the filename **Third web page with a stylesheet.htm**

3 Load your browser and load the file **Third web page with a stylesheet.htm**

Check carefully that the stylesheet is producing the correct styles for the web page.

Hyperlinks (often just called links)

Hyperlinks (links) are an important part of web sites as they allow users to move from one web page to another or within the same web page. Links are text or images that a user can click on in order to jump to a new web page, which can be on the same web site or a completely different one. They can also be used to move between sections on the same web page which is useful if the web page is long. When the cursor is moved over a link, the arrow turns into a small hand.

Creating external links to a web site using the URL

You can create HTML code that will provide a link to an external web site using the URL (i.e., web address) of the external site. External links are used to provide links to web pages and web sites that are outside your directory where you store your web site.

For example, if we wanted to provide a link to the Oxford University Press web site the HTML for this is:

```
<a href="http://www.oup.com/">Oxford University Press web site</a>
```

The text shown in blue is the URL of the web site you want to link to and the text to supply the link (i.e., Oxford University Press) also appears in the tag.

The link on the web page appears as follows: Oxford University Press web site

Important note

To create a link to an external web site you have to use http:// as the starting part of the URL and not just www.

It is also important to check external links regularly as web sites sometimes change their URL, meaning the link to them will no longer work.

Creating a link in a new window

Unless specified, a link will open in the same window.

To create the HTML to open a link in a new window you add **target="_blank"** to the anchor tag as shown here:

```
<a href="http://www.oup.com/"target="_blank">Oxford University Press web site</a>
```

This results in the link appearing like this on the web page: Oxford University Press web site

When the user clicks on the link they are taken to the web site with URL **http://www.oup.com/** opened in a new window.

Creating links to other locally stored web pages

Web sites usually consist of many web pages and users will want to be free to move between these pages using links. These links are called relative links and are used to enable users to move to other pages that are saved in the same directory on the computer or server.

The HTML you will need to enter in the editor for this feature looks like this:

```
<a href="webpage1.html">Introduction to HTML</a>
```

The text shown in blue is the name of the file for the web page you are linking to and after it is the text that appears on the web page and acts as the link. When viewed in the browser the link appears as follows:

Introduction to HTML

Creating links to send mail to a specified email address

Visitors to your web site may want to send you a message and you can create a link that will allow them to do this using the mailto tag. When the user of your web page clicks on the text underlined acting as the link, a dialog box addressed to you will appear into which they can type their message.

The HTML you will need to enter in the editor for this feature looks like this:

<a href=**mailto:emailaddress@aol.com**>Name to go here</a>

The text shown in blue is your email address and after it is the name that appears on the web page that acts as the link. When viewed in the browser, the link appears as follows: Name to go here

Creating anchors and links to anchors on the same web page

Anchors act as points of reference on a web page. They enable a user to move to part of the same web page usually using an item in a menu as a link. It makes it easier for a user because they do not have to waste time looking at part of the web page they are not interested in.

Suppose there is a menu like this at the start of the web page and then material about each of these in turn is discussed in sections on the same web page:

1. What is copyright?
2. When can you use copyrighted material?
3. Myths about copyright

To create the above links to the anchors in the main body of text in the document, you can use the following HTML:

<a href="#copyright">1. What is copyright?</a>
<a href="#use">2. When can you use copyrighted material?</a>
<a href="#myths">3. Myths about copyright</a>

The above links will link to the anchors that are positioned in the text to which they refer. So, for example, the link called "copyright" will need to have an anchor called "copyright" at the start of the "What is copyright?" section. Usually you would place the anchor near the heading for the section.

The HTML used to provide an anchor for the first item in the list could be:

<a name="copyright"></a>

The HTML for the second item in the list could be:

<a name="use"></a>

The HTML for the third item in the list could be:

<a name="myths"></a>

Using tables to organize a web page

Planning out a web page is difficult but it can be made easier by using a table. The table produces a page layout into which text, images and other elements can be placed. Tables are also used in web pages for presenting content in a similar way to the way tables are used in ordinary documents.

In the examination you may be asked to create a table consisting of a certain number of columns and rows. In many cases there will be a diagram to show what the table should look like on the web page.

There are three basic tags used to create a table:

<table> this is the tag to start a table
   <tr> this is the tag to start a row of a table
      <td> this is the tag to start a data cell
      </td>
   </tr>
</table>

Here is a section of HTML used to create a table:

<table>
   <tr>
      <td>Cell 1</td>
      <td>Cell 2</td>
   </tr>
   <tr>
      <td>Cell 3</td>
      <td>Cell 4</td>
   </tr>
</table>

The table for this section looks like this:

Cell 1 Cell 2
Cell 3 Cell 4

Notice that there is a <tr> and </tr> for each row in the table. Also there is a <td> and </td> for each column in the row (or each cell as the intersection of a column and a row is called a cell).

Notice that there is no border to the table. To include a border, a border attribute has to be used as you will see later.

Borders around tables

The tables so far did not have borders around them. To place a border around the table you use the following:

<table border="2">

Activity 15.5

Creating anchor links

In this activity you will learn the following skills:

▸▸ Create anchor links in a web page

Sometimes anchor points take the user to another point on a long web page and, to avoid the user having to use the scroll bar to scroll the page back to the beginning, it is best to supply them with a link. This is done by inserting an anchor at the top of the page and then placing a link further down the page. You can see the link at the bottom of the page in the following screenshot:

Myths about copyright

Here are some myths about copyright:

There is no copyright symbol, so it is ok to use

Wrong. As soon as a piece of work is produced it is copyrighted automatically with or without the copyright symbol.

I have acknowledged the source of the material so it is ok to copy it

Not true. You still should ask permission.

It is on a website, so it is ok to copy it

Wrong. Someone has spent time and money producing the item you are copying. Copying it is wrong and illegal

How do I know whether work I wish to copy is copyright free?

You have to assume that no work is copyright free unless it specifically says so. Even then you need to check the conditions carefully.

How do I copyright the original work that I produce?

Copyright material is sometimes marked with a © next to it or near it. If you stress that the material is copyright, it is a good idea to include this symbol in the following way:

© Author or owner of copyright. Date of publication

© Stephen Doyle. 2012

It is not a legal requirement to put the symbol in, because any original work is automatically given copyright.

Copyright when taking your own photographs

If possible, it is often quicker to go out and take your own photographs because there are no copyright problems. You own the copyright on photographs you take yourself. There are cases where you would have to use other photographs taken by others. An example of this would be if you wanted a photograph of a distant landmark such as the Great Wall of China or someone famous such as the President of the USA. There are many photograph libraries on the Internet where you can obtain photographs and use them on your website for a small fee.

Go back to the top

1 Load Notepad and open the HTML file called **Anchors**.

2 Look carefully at the HTML and notice the construction of the anchor link which takes a user from the bottom of the page to the top.

3 Use this document to create appropriate anchor links of your own.

Activity 15.6

Creating the code for a table

In this activity you will learn the following skills:

▸▸ How to create a basic table

Write the section of HTML that can be used to produce a table with three columns and four rows. Each cell in the table should hold the following text.

Cell 1 Cell 2 Cell 3
Cell 4 Cell 5 Cell 6
Cell 7 Cell 8 Cell 9

This needs to be added after the <table> and before the instructions which specify the rows and cells.

The number "2" is used to specify the thickness of the border.

Here is a table that contains a border of thickness 2 pixels:

Cell 1	Cell 2	Cell 3
Cell 4	Cell 5	Cell 6
Cell 7	Cell 8	Cell 9

Changing the width of a table
By adding the following, when the table is drawn it will have a border of 1 pixel and the whole table will occupy a width of 50% of the entire screen width.

```
<table border ="1" width="50%">
```

In the browser the table will appear like this:

Cell 1	Cell 2	Cell 3
Cell 4	Cell 5	Cell 6
Cell 7	Cell 8	Cell 9

The table width can also be specified by the number of pixels like this:

```
<table border ="1" width="500">
```

Changing the height of a table
The following HTML will set the height of a table to 250 pixels and the width of the table to 300 pixels. The border command ensures that a border is drawn around the table:

```
<table height="250" width="300" border="2">
```

Other attributes of tables
Aligning the text horizontally within a cell
The contents of cells can be aligned horizontally left, centred or right using the following:

```
<td align="left"></td>
<td align="center"></td>
<td align="right"></td>
```

These are used like this with text:

```
<td align="right">This text will be aligned to the right</td>
```

Using merged cells
If you want to merge cells in a table together there are two attributes you can use depending on whether you want to merge cells in columns or rows. These attributes are:

▸▸ Colspan
▸▸ Rowspan

Colspan
Colspan is an attribute using in the <td> tag and you can use it to specify how many columns the cell should span. Look at the following section of HTML:

```
<table border ="2">
    <tr>
        <td colspan="3">Cell 1</td>
    </tr>
    <tr>
        <td>Cell 2</td>
        <td>Cell 3</td>
        <td>Cell 4</td>
    </tr>
</table>
```

You can tell that this table consists of two rows (as there are two <tr> tags). You can also tell that the second row is divided into 3 cells (as there are three <td> tags (one for each cell)). There is a border around the table of 2 pixels.

Colspan is short for column span.

```
<td colspan="3">Cell 1</td>
```

The above HTML tells us the first row spans 3 columns and contains the text "Cell 1".

This section of code produces the following table:

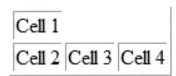

Rowspan
Rowspan specifies how many rows a cell should span over.

Aligning a table horizontally on the screen
You may be asked in the examination to align the table centrally in a horizontal direction on the screen. To do this, you replace the <table> </table> commands above at the start and end of the table with the following:

```
<table align ="center">
</table>
```

Cell spacing and cell padding

To add spacing between the cells in a table you use the cellspacing attribute. To add spaces between the content of a cell and the cell itself you use the cellpadding attribute. You will see how both of these attributes can be used in the following section.

Changing the width of a cell

The width of a cell can be set like this:

```
<td width="20%">Cell 1</td>
```

This changes the width of the cell to 20% of the table width.

The width of a cell can also be specified in pixels like this:

```
<td width="200">Cell 1</td>
```

The cell here has a width of 200 pixels.

Activity 15.7

Changing the width and height of a table and the width of cells in the table

In this activity you will learn the following skills:

▶▶ How to change the width and height of a table
▶▶ How to change the width of cells in a table

1   Load your HTML editor (e.g., Notepad) and enter the following HTML code

```
<table height="250" width="300" border="2">
    <tr>
            <td width="200">Cell 1</td>
            <td width="50">Cell 2</td>
            <td width="50">Cell 3</td>
    </tr>
    <tr>
            <td width="200">Cell 4</td>
            <td width="50">Cell 5</td>
            <td width="50">Cell 6</td>
    </tr>
</table>
```

Notice how the height and width of the table in pixels are specified in the first line.

Notice also how the width of each cell is specified in pixels.

2   Save this HTML in your HTML folder using the filename **Table Activity 15.7**

Open this file in your browser to check it correctly produces the basic table.

Check your table looks like this:

Cell 1	Cell 2	Cell 3
Cell 4	Cell 5	Cell 6

Table headers

Table headers can be used to specify the header of a column. In the following table "Country", "Capital" and "Currency" have been used as the headers.

Country	Capital	Currency
Greece	Athens	Euro

Here is the HTML used to create the above table. Notice that the table width is 60% of the screen width and the cell spacing and cell padding have both been set to 5 pixels.

```
<table width="60%" border="1" cellspacing="5"
cellpadding="5">
    <tr>
        <th>Country</th>
        <th>Capital</th>
        <th>Currency</th>
    </tr>
    <tr>
        <td>Greece</td>
        <td>Athens</td>
        <td>Euro</td>
    </tr>
</table>
```

Activity 15.8

Border, cell spacing and cell padding

In this activity you will learn the following skills:

▸▸ How to insert a table using a table header, table rows, table data, cell spacing, and cell padding

1 Load your HTML editor (e.g., Notepad) and enter the code to produce the Country, Capital and Currency table shown above but alter the first line to the following:

```
<table width="60%" border="1" cellspacing="10" cellpadding="50">
```

Your table should look the same as this:

Country	Capital	Currency
Greece	Athens	Euro

2 Now amend the HTML in the following way:

Set the width of the table to 80%

Set the table border to 1

Set the cellspacing to 1

Set the cellpadding to 10

Save the file using a suitable filename and then load it into your browser.

Your table should appear like this:

Country	Capital	Currency
Greece	Athens	Euro

3 Add another row to the table containing the text: France, Paris and Euro.

4 Save the file using a different filename to the name given in step 2 and then load it into your browser.

Activity 15.9

Producing tables

In this activity you will learn the following skills:

▸▸ Use tables to organize a web page

Being able to produce tables according to a design is needed for the examination. Here are some tables that you need to work through and understand. Study the code carefully and then type them in and save each using the filenames **Example Table1.htm, Example Table2.htm, etc.**

Table 1

This table has a horizontal width of 60%. Notice that the width of a cell can be specified and this width is a percentage of the table size and not the screen size.

```
<table border="1" cellpadding="5" cellspacing="5" width="60%">

  <tr>

    <td width="20%">Cell 1</td><td>Cell 2</td>

  </tr>

  <tr>

    <td width="20%">Cell 3</td><td>Cell 4</td>

  </tr>

</table>
```

Cell 1	Cell 2
Cell 3	Cell 4

Table 2

This table uses a header that spans two columns:

```
<table border="1" cellpadding="5" cellspacing="5" width="50%">

  <tr>

    <th colspan="2">Table header</th>

  </tr>

  <tr>

    <td width="30%">Cell 1</td><td>Cell 2</td>

  </tr>

</table>
```

Table header	
Cell 1	Cell 2

Table 3

Notice that in the first row of the table, the first cell occupies two rows and the second cell occupies the normal one row. In the second row the cell is placed in the space left to the right of cell 1.

```
<table border="1" cellpadding="5" cellspacing="5" width="50%" >

  <tr>

    <td rowspan="2">Cell 1</td><td>Cell 2</td>

  </tr>

  <tr>

    <td>Cell 3</td>

  </tr>

</table>
```

➔

| Cell 1 | Cell 2 |
| | Cell 3 |

Table 4

The first row of this table is used as a header and spans two columns. The second row contains Cell 1 which spans two rows and Cell 2 which spans a single row. The third row only contains Cell 3. The fourth row contains Cell 4 which spans two columns.

```
<table border="1" cellpadding="5" cellspacing="5" width="50%" >
    <tr>
        <th colspan="2">Table Header</th>
    </tr>
    <tr>
        <td rowspan="2">Cell 1</td><td>Cell 2</td>
    </tr>
    <tr>
        <td>Cell 3</td>
    </tr>
    <tr>
        <td colspan="2">Cell 4</td>
    </tr>
</table>
```

Table Header	
Cell 1	Cell 2
	Cell 3
Cell 4	

Activity 15.10

Producing tables to a design

In this activity you will learn the following skills:

▸▸ Use tables to organize a web page

In the examination you may be given a diagram of a table and then have to produce the section of HTML needed to create it.

Produce the HTML code for each of the following tables.

Table 1

This table is to have a width of 50% of the screen, a table border of 1, cell spacing of 5 and cell padding of 20.

The table is to appear like this:

| Cell 1 | Cell 2 | |
| Cell 3 | Cell 4 | Cell 5 |

Table 2

This table is to have a width of 800 pixels, a table border of 1, cell spacing of 5 and cell padding of 20. Cell 2 is to span two rows.

The table is to appear like this:

Cell 1	Cell 2
Cell 3	
Cell 4	Cell 5

Table 3

This table is to have a width of 500 pixels and a table border of 1. Cell 1 is to span three rows.

The table is to appear like this:

Cell 1	Cell 2
	Cell 3
	Cell 4

Changing the background colour of a table, row or cell

The background colour of a table, row or cell can be changed. You can either use the name of the colour or pick the colour from a colour chart containing the hexadecimal code for the colour.

This section of HTML is used to change the background colour of a row to red:

```
<table border="1" cellpadding="10" cellspacing="5">
    <tr bgcolor="#FF0000">
        <td>Cell 1</td><td>Cell 2</td>
    </tr>
    <tr>
        <td>Cell 3</td><td>Cell 4</td>
    </tr>
</table>
```

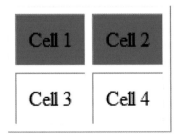

Activity 15.11

Changing background colours in tables

In this activity you will learn the following skills:

▸▸ Change the background colour of a row, cell or table

1   Enter the HTML to draw the table shown on the left into Notepad.

2   Save the file using the filename **bgcolour1.htm** and load the file into the browser to check that it produces the table shown on the left.

3   Go back to the HTML and select the text bgcolor="#FF0000" by highlighting it and cut it and paste it into the position shown here:

```
<table border="1" cellpadding="10" cellspacing="5">
    <tr>
        <td bgcolor="#FF0000">Cell 1</td><td>Cell 2</td>
    </tr>
    <tr>
        <td>Cell 3</td><td>Cell 4</td>
    </tr>
</table>
```

The background colour is now applied to a single cell (i.e., Cell 1).

Save the file using the filename **bgcolour2.htm**

Load it into the browser to check it is producing the correct result.

4   Go back to the HTML and then make your own adjustment to make the whole table have a red background.

Save the file using the filename **bgcolour3.htm** and check that it works correctly.

Using images on web pages

There are many different file formats for images and not all of them work with all browsers. It is therefore best to use images in the following formats:

▸▸ .jpg (digital cameras store images in this format)

▸▸ .gif (often used for simple line diagrams and clip art).

In order to use an image on a web page, the image file must be stored in the same folder as the web page. This means the images you intend to use must be copied to the same folder.

Adding an image to a web page

Depending on the file format for the image, this is done using either:

    <img src="filenameofyourimage.jpg"/>

or

    <img src="filenameofyourimage.gif"/>

Finding out file format and size of an image

It is important to know the file format and size of an image. It is easy to do this when the image is displayed by displaying the image and then right clicking on the mouse so the following menu appears:

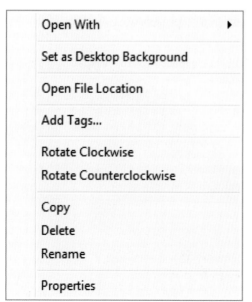

Click on Properties and the following information about the image will be displayed:

The information we require is the dimensions of the image. This image is 2304 pixels wide and 1728 pixels high. If you scroll further down you will see the file format and also the size of the file.

Images taken with a digital camera are far too large to be used on a web page. They need to be re-sized and saved using a different filename before being used on a web page. To re-size an image you need to use a photo-editing package such as Photoshop or Fireworks. Here the image size has been altered to 200 pixels wide and 150 pixels high.

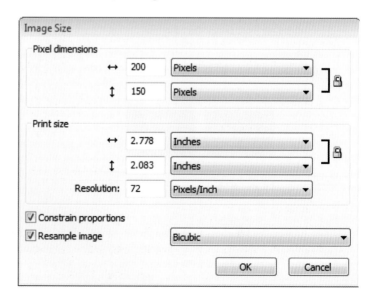

Using tags to adjust image size

The height and width attributes are used to set the height and width of an image on a web page. Using these you can specify how big the image appears on the web page regardless of how large the image actually is.

If you do not set a height and width for the image, it will appear in the actual size it was saved.

```
<img src="filenameofyourimage.jpg" width="100"
height="100"/>
```

This displays the image whose filename is "filenameofyourimage" using a width of 100 pixels and a height of 100 pixels.

Although you may think that this reduces the file size of the image and hence the time taken to load the image onto the web page, this is not the case, as the image is still loaded from the disk and then adjusted to fit the size specified.

It is therefore better to reduce the image size as suggested before using an image editing program, as this will reduce the file size when the adjusted image is saved. This will make the file size for the image smaller and hence faster to load on the web page.

It is still useful to use width and height attributes because it tells the browser software how much space is needed in the final layout and this makes the web page download faster.

Positioning of images and text in tables (vertical and horizontal alignment)

Once a table structure has been defined, content such as text and images can be added to the cells. How an image appears in a cell depends on the alignment used for that cell. Both text and images can be aligned vertically in a cell using the following:

```
<td valign="top"><This text will appear aligned to the top of
the cell></td>
```

The above valign will set the vertical alignment of the image to the top of the cell.

```
<td valign="top"><img src="picture1.jpg"></td>
```

The above valign will set the vertical alignment of the image in the cell to the top.

You can have the following vertical alignment:

valign ="top"

valign ="middle"

valign ="bottom"

For horizontal alignment of text and images, the following is used:

```
<td align="center"><This text will be aligned centrally></td>
```

The above HTML aligns the text centrally in the cell.

```
<td align="right"><img src="picture1.jpg"></td>
```

The above HTML aligns the text to the right in the cell.

You can have the following vertical alignment:

align = "left"

align ="center"

align ="right"

Activity 15.12

Inserting an image into a cell in a table

In this activity you will learn the following skills:

▸▸ Insert an image into a cell in a table

▸▸ Align an image horizontally and vertically in a cell

1 Before starting this activity you will need to check that the image file called **Oxford_banner.jpg** appears in the same folder that you use to keep your HTML files. If the file is not in the folder, you need to copy it from the CD into the folder.

2 Load Notepad and open the file called **Table with image.htm**

The following HTML appears:

```
<table border="1" cellspacing="5" cellpadding="5" width="50%">
        <tr>
                <td><img src="Oxford_banner.jpg" width="600" height="200"></td><td>Cell 2</td>
        </tr>
        <tr>
                <td>Cell 3</td><td>Cell 4</td>
        </tr>
</table>
```

Notice the way the tag for the image has been inserted into the first cell of the table. The filename of the image to be inserted is between quotation marks and also the width and height of the image have been specified in pixels.

```
<img src="Oxford_banner.jpg" width="600" height="200">
```

➡

3 Load the file **Table with image.htm** into the browser and you will see the image appear in the table like this:

	Cell 2
Cell 3	Cell 4

4 There are three images you now need to place in the other cells. Each image is to have a width of 200 pixels and a height of 100 pixels. The image files are to be placed as follows:

In Cell 2: oup1

In Cell 3: oup2

In Cell 4: oup3

Adjust the HTML by replacing the text Cell 2, Cell 3 and Cell 4 with the above images. Check that when viewed in the Browser the web page looks like this:

5 You are now going to alter the alignment of the content of the cells so that the image in the top right cell is aligned to the top and the image in the bottom left cell is aligned to the right.

Change the tag for the top right data cell to:

    <td valign="top"><img src="oup1.jpg" width="200" height="100"></td>

Change the tag for the bottom left data cell to:

    <td align="right"><img src="oup2.jpg" width="200" height="100"></td>

Save the file using the file name **Table with image1.htm**

Load the file into your browser and check it appears as in the screenshot above.

Adjusting the alignment of the image on a web page

You can adjust the alignment of the image on a web page. This section of HTML aligns the image to the middle of the page as well as setting the height and width of the image.

    <img src="filenameofyourimage.jpg" align="middle" width="100" height="100"/>

Here is a list showing the alignments you can have:

    align="middle"

    align="left"

    align="right"

    align="top"

    align="bottom"

All these alignments need to be inserted into the tag for the image similar to that shown above.

For example:

    <img src="image.jpg" align="left"/>

Aligns the image having the filename image.jpg to the left of the page.

Using software to resize an image

It is often necessary to resize an image so that it can be put into a certain space on a web page. There are two ways this can be done:

Use image editing software (e.g., Adobe Photoshop, Microsoft Photo Editor, Macromedia Fireworks, etc.)

Specify the required height and width of the image in pixels in the image tag like this:

    <img src="oup1.jpg" width="200" height="100">

Maintaining the aspect ratio

Maintaining the aspect ratio means keeping the ratio of the width to the height of an image the same when the image is resized. This can be done in the image tag as follows:

    <img src="oup1.jpg" width="200">

The image in the image tag is resized so that it has a width of 200 pixels. Notice that the height is not specified in the tag as it will be automatically set to maintain the aspect ratio.

Distorting an image

Sometimes an image needs to be stretched in one direction only. For example, the shape of the original image is not quite the same shape as the gap on the web page it needs to fill. Stretching an image in one direction means that the aspect ratio is not being maintained.

The top image is the original and the bottom image is distorted because it has been widened while keeping the height the same as the first image. You can see that the image has been stretched in the horizontal direction only.

Aspect ratio

The aspect ratio of an image is the ratio of the width of the image to its height. It is expressed as two numbers separated by a colon. Suppose the aspect ratio of an original image is 2:1, you can distort the image using a tag by specifying the width and height like this:

```
<img src="oup1.jpg" width="200" height ="120">
```

Notice that the ratio of the width to the height is no longer 2:1 so the aspect ratio has changed meaning that the image has been distorted slightly.

Using images as links

Links can be made using images. When the user clicks on the image used as the link they will be taken to a new web page or web site. It is important to make it clear to a user that the image is acting as a link and that they have to click on the image to follow the link. This can be done by putting a short message next to the image.

```
<html>
    <h1>International resources</h1>
    <p align="center"><a href="index.htm"><img src="oup1.jpg"
    width="200" height="200" border="3" alt="Click here for
    the link to international resources"</a></p>
    <p align="center">Click on the image to follow the link</p>
</html>
```

Look carefully at the above section of HTML. The text "International resources" will appear at the top of the page as a heading.

A new paragraph will be started and an image will be centred in the horizontal direction and the image with the filename "oup1. jpg" will be sized and placed with a border in this position.

This image also acts as a link to the locally stored web page with the file name "index.htm" and when the user puts the cursor over the image, the text "Click here for the link to international resources" will appear.

Look at the following line in the above code:

```
<p align="center">Click on the image to follow the link</p>
```

This will display the message "Click on the image to follow the link" which is positioned centrally and under the image like this:

Click on the image to follow the link

The need for low resolution images for data transfer

When an image is used on a website, the user of the website has to wait while the image is obtained from the server, transferred over the Internet and assembled on the web page using the browser software. The time they have to wait for the image file to be transferred is called the data transfer time (also called the download time). If the user has a high speed broadband Internet connection this time will be low. If they have a low speed connection there will be a time delay and this is annoying. The time for the image to appear depends on the file size of the image. It is therefore advisable to keep image files as small as possible.

For the above reason, images in web sites are normally used at low resolutions as low resolution images have a smaller file size.

Amending the file type

There are many file types for images but only three that are used with web sites:

▸▸ GIF file type images only consist of 256 colours this file type is good at compressing images so that they load quickly.

▸▸ JPEG/JPG file type images comprise millions of colours and this makes them the main file type for photographic images.

▸▸ PNG file type images are able to use millions of colours and are good at compressing images.

All graphics packages such as Fireworks or Photoshop are able to change different file types to those shown above. All you need to do is to select Save As... and select the file type you want to save in.

Printing web sites in browser view

Printing web sites in browser view means printing the web pages exactly as they appear when viewed in the browser you use. In the examination, you will usually be asked to add your name and candidate number on the web pages so that when they are printed, your work can be identified.

Printing evidence of HTML code or stylesheets

If you are asked to supply a copy of the HTML code or stylesheet or both, you will be asked to supply a printout containing details such as your name, candidate number and centre number. As always, you must follow the exact instructions given about the details required in the examination paper.

To produce a printout containing these details it is best to take a screenshot of the HTML or stylesheet using the Prt Scr key on the keyboard. This will take a copy of the screen and store it in the clipboard. The screenshot can then be pasted into a new word-processing document where it can be cropped to remove any unwanted parts and re-sized so that all information is clearly seen. The details such as name, candidate number and centre number can then be added before saving the document and then printing it out.

You can also print the HTML code directly from some editors. Print Screens may not show enough for a long/wide string of code.

> **Practical exam tip** All documents and printouts must have your name and some other details printed (not written) on them.
>
> You might also have to print the CSS file – again putting your name on it. You can use Print screen and copy the screen to word-processing software such as MS Word and then type the details such as your name and other details before printing out the document.

16 Presentation authoring

You will probably have created many presentations in different subject areas but you may not have developed skills in all the areas needed for the examination. For the examination you need to have the skills to be able to create and control an interactive presentation.

In this chapter you will be learning how to create presentations using the software MS PowerPoint 2007. There is other presentation software available having similar features that could be used. MS PowerPoint is a brand name so you would get no marks for giving brand names in the examination. So rather than say "use MS PowerPoint" you would say use "presentation software".

The key concepts covered in this chapter are:
▶▶ Using a master slide to place objects and set styles
▶▶ Creating presentation slides
▶▶ Creating notes for the presenter and audience

Creating a presentation

The need for consistency in a presentation

All the slides in a presentation should look as if they belong together. This means they should all have a similar design. It is acceptable that the first slide is different to the others because it usually shows just the title of the presentation and sometimes the presenter's name.

Consistency needs to be considered before you start your presentation. Here are some of the things you will need to consider:

▶▶ Slide design – slides need to have a consistent design that has been designed in advance. Once the presentation has been produced it needs to be checked again to ensure that the design is consistent.
▶▶ Fonts – you need to choose a set of fonts that work well together. Fonts need to be consistent from one slide to the next.
▶▶ Point sizes – these are the size of the font. You will need to choose what point size for headings, sub-headings, and body text (i.e., the main text).
▶▶ Colour schemes – all the slides need to use the same combination of colours for components such as banners, borders, etc.
▶▶ Transitions – these are the movement from one slide to the next. There are lots of eye catching transitions but you need to make sure you do not use a different one for each slide transition.
▶▶ Animations – animations are movement on the actual slide. For example, you can animate the way bullet points appear on each slide. Make sure that animations are used consistently.

In the examination you will be usually given instructions about the above so you need to understand how to create/change them using the presentation software.

What is a master slide and why are they important?

Master slides are used to help ensure consistency from slide to slide in a presentation. Master slides (also called slide masters) are used to place objects and set styles on each slide. Using master slides you can format titles, backgrounds, colour schemes, dates, slide numbers, etc.

In the examination you will be given details of designs, colour schemes, fonts, etc., that you must add to a master slide.

The footer, date and slide number areas on the master slide

There is an area at the bottom of the slide master which is used to add certain information at the bottom of every slide in the presentation, such as a footer, page date, or slide number. The footer can be used for text such as a copyright message (e.g., ©Stephen Doyle 2011).

If you want to add text in the footer area place holder (i.e., the rectangle with footer in it) you click in the box and type in the text.

The date and slide number (i.e., #) are already on the slide master but they will not be shown on the actual slides unless you make them active. To make these details appear on each page you have to follow the following steps:

1   Click on [Insert] and then select Header & Footer and the following dialogue box appears where you can enter dates, slide numbers, and a footer.

![Header and Footer dialogue box showing Slide tab with Include on slide options: Date and time (Update automatically 30/11/2010, Language: English (United Kingdom), Calendar type: Western, Fixed 30/11/2010), Slide number, Footer, Don't show on title slide. Buttons: Apply to All, Apply, Cancel, and Preview.]

If you want a date and time put on each slide make sure that you put a tick in the box. You are then presented with a list of formats for the date to choose from:

If you want the slide to always show the current date, then you should select Update automatically like this:

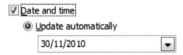

Sometimes you will want to show the same date on each slide such as the date you produced the presentation.

◉ Fi̲xed

If you select 30/11/2010 this will mean that a certain date will always appear on each slide.

2 To put a slide number on each slide, ensure that slide number has been selected like this:

3 You can just enter the footer text into the placeholder area marked "Footer" on the master slide or you can enter the text for the footer into the Header and Footer dialogue box like this:

Note the (c) next to the name will automatically turn into a copyright symbol on the slides. Remember you can put any text into the footer and will be asked to do so for the examination.

Important information about headers and footers

You will see the footer placeholder areas at the bottom of the slide. If you do not require any of these, you do not have to delete them. They will only appear if you set them to appear using the method shown above.

If a footer place holder is not in the position asked for in the examination then you can move it. You can also enlarge it so that it occupies the entire width of the slide.

Activity 16.1

Creating a master slide

In this activity you will learn the following skills:

▸▸ How to create a master slide which can be used to ensure the consistency of design in all the other slides in the presentation

▸▸ How to change the background colour of a slide

▸▸ How to add a text box to a slide

▸▸ How to add an image to a slide

▸▸ How to add text to the header and footer

1 Load the presentation software Microsoft PowerPoint.

2 Click on the View tab View and select Slide Master .

3 The following screen appears:

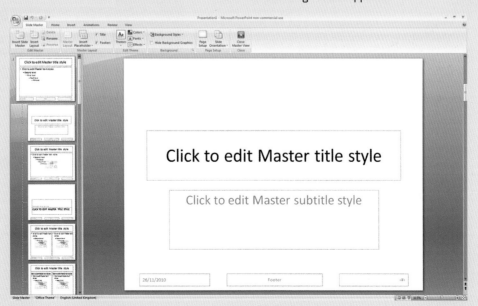

4 You now have to consider the medium for your presentation. In this case it is a presentation on a screen in landscape orientation with the audience notes (notes accompanying the presentation given to the audience) and the presenter notes (note giving hints when giving the presentation).

Click on Page Setup .

This allows you to set the size of the screen you are displaying the presentation on and the orientation of slides and notes.

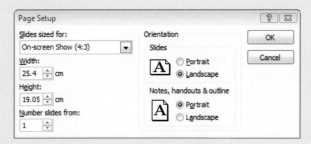

The default settings (settings that will be used unless you change them) are used for this presentation. Confirm by clicking on OK.

5 Click on the first slide at the top of the list of slides. This is the Master Slide and you can use this one slide to layout the design of your slides as objects such as text and images (clip art, logos, lines, shapes, etc.) will be applied to all the slides in the presentation. This means that if you inserted a logo on this slide, it will appear at the same position on all the slides in the same presentation.

The other slides you see under this one contain different style slides and these can be chosen according to the content you want to put on the slide.

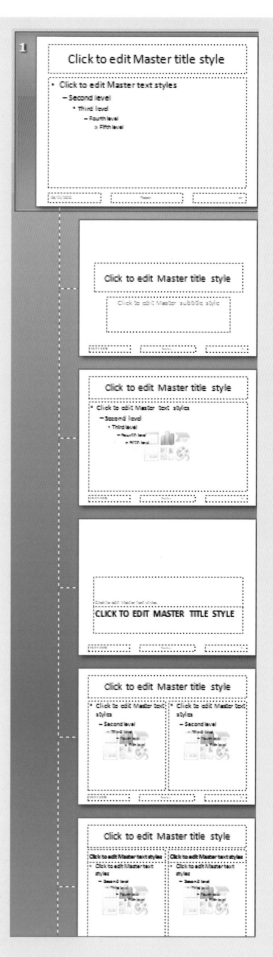

6 You are going to set up a master slide:

Ensure you have selected the master slide shown here:

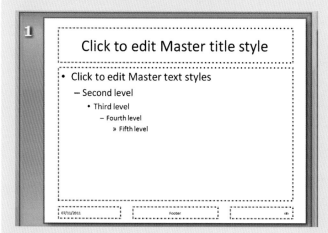

The following master slide is shown.

You are going to change the background colour of all the slides from white to a pale yellow.

Right click anywhere on the white part of the slide and the following menu is displayed:

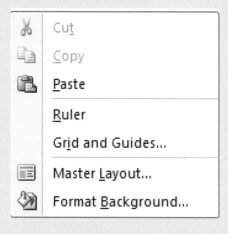

Left click on Format Background and the following window is displayed:

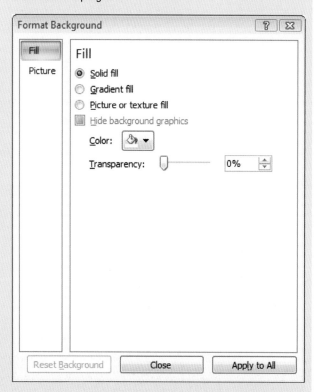

Check that Solid fill has been selected and then click on the drop-down arrow for the colour 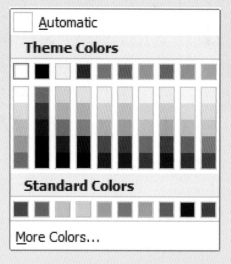 and the following palette appears:

Click on the bright yellow colour in the Standard Colours section. Notice that all the slides now have this colour as the background. Now adjust the transparency of the colour by setting it to 80%:

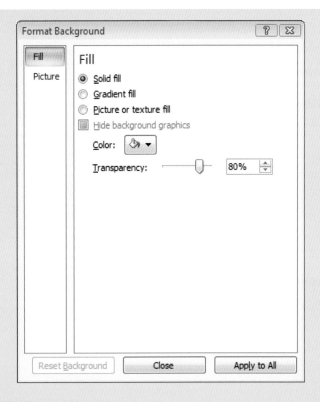

Click on Close.

7 You are now going to add a text box to the master slide containing the following details:

▸▸ Your full name
▸▸ Your centre number (use 12122 if you don't have this information)
▸▸ Your candidate number (use 123123 if you don't have this information)

These details need to be in a sans serif font and have a point size of 12 Pt and be positioned in the top left-hand corner of the slide.

There is already an object (i.e., a box containing the text "Click to edit Master title style") and this needs to be resized to make room for the text box we want to insert.

Click on the corner like this and use the handle and drag the corner down and to the right:

Click to edit Master title style

The box should be moved into a position like this:

Click to edit Master title style

Click on the insert tab Insert and then select Text Box Text Box .

Now draw the text box in a similar position to that shown here:

Change the font to Arial (which is a sans serif font) and the font size to 12 Pt by altering the settings like this:

Type the following information into the text box:

Your name 12122   123123

Your text box containing the information will now look like this:

Stephen Doyle   12122   123123

Click on a blank area of the slide to deselect the text box.

As the text in the text box is on the master slide, it will appear in the same position on all the slides.

8 You are now going to insert a piece of clip art on the master slide positioned in the bottom right-hand corner of the slide.

The piece of clip art needs to be a picture of a car (any car will do).

Click on Insert and then on Clip Art and the following clip art pane appears:

Click on "Car" in the search box and then click on Go .

Find a suitable picture of a car and left click on it and then, keeping the left mouse button pressed down, drag onto the position on the slide and, when in the correct position, release the mouse button. The piece of clip art will then appear in position like this:

Stephen Doyle 12122 123123

Click to edit
Master title style

- Click to edit Master text styles
  - Second level
    - Third level
      - Fourth level
        » Fifth level

28/11/2010   Footer   ‹#›

Again this image will appear in an identical position on all the slides in the presentation.

If any of these boxes are needed but not in the position shown, you can select the box and drag it into a new position.

9   You are now going to re-position the box containing the slide number so that it is on the left. You will need to move the date out of the way while you do this by left clicking on it and drag it to another position. Do not worry about putting it into the final position as you can do this later.

29/11/2010

Footer   ‹#›

Left click on the box with the slide number in it (i.e., shown as #) and keeping the mouse button pressed down, drag the box over to the left-hand side of the screen. Now in a similar way move the box containing the date into the position shown below:

‹#›   Footer   29/11/2010

The slide number is right aligned in the box.

You are required to change this to left alignment and to do this click on  and then on ▤.

The slide number will move to its new position like this:

The slide number will now appear on each slide positioned in the bottom left-hand corner.

10 Details such as slide number, the footer, and the date need to be turned on before they will be shown on the slides.

Click on Insert and then select Header & Footer and the following dialogue box appears:

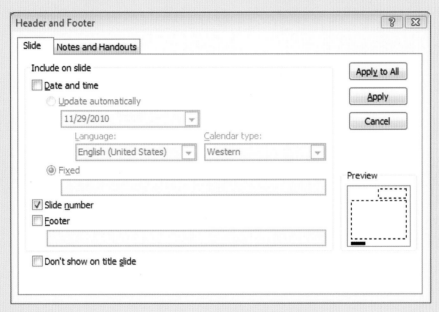

Click on Apply to All.

Ensure that there is a tick on the Slide number box. Notice also that you could turn the date on and put some text in the footer if asked to. Notice the preview shows that only the slide number will be shown.

11 You now need to check that the objects you have placed on the master slide do not interfere with any of the boxes into which you type the content when adding the material to the slides.

Click on the second slide in the list on the left.

You can see that the clip art overlaps the box to hold the subtitle:

Stephen Doyle 12122 123123

Click to edit Master title style

Click to edit Master subtitle style

‹#›   Footer   29/11/2010

To solve this problem we can re-position the two boxes by moving them up slightly. To move them you need to left click on the border of the box and then keeping the left mouse button pressed down, drag them into a position similar to that shown here:

Stephen Doyle 12122 123123

Click to edit Master title style

Click to edit Master subtitle style

12 You now have to set up the following styles for the presentation:

▸▸ Heading: dark blue, 48 point, right aligned serif font.

▸▸ Subheading: black, centre aligned, 24 point serif font.

▸▸ Bulleted list: black, left aligned 18 point sans serif font and you are able to choose the style of the bullet used.

Click on the text "Click to edit Master title style" as this is the heading for the slide.

Choose any serif font (e.g., Times New Roman) and change the font size to 48 like this:

| Times New Roma ▾ | 48 ▾ |

Click on [A ▾] and select dark blue (it does not matter which dark blue as there are several) from the colour chart:

Click on [≡] to right align the heading.

Click on the text "Click to edit Master subtitle style" as this is the subheading for the slide.

Choose any serif font (e.g., Times New Roman) and change the font size to 24 like this:

| Times New Roma ▾ | 24 ▾ |

Click on [A ▾] and from the colour chart select the colour black.

Notice that there are no bullet points on this first slide.

Click on one of the slides whose layout contains bullet points and then click on the first level of bullet points (i.e., the first bullet point on the master slide):

- ## Click to edit Master text styles

Click on the text next to the bullet point and choose a sans serif font (e.g., Courier, Calibri, etc.) and set the font size to 18. Set the colour of the font to black and set the text to be left aligned.

Change the shape of the bullet by clicking on the drop down arrow [≣ ▾] and choosing a bullet shape of your choice by clicking on it like this:

⇥

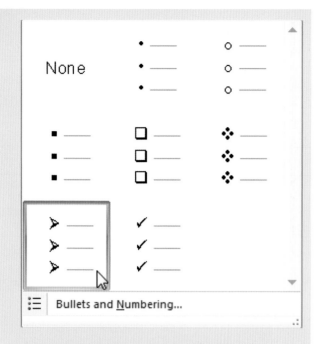

You have now made your first master slide.

Save this master slide using the filename **Master**.

Creating a master slide that includes shaded areas and lines

In this activity you will learn the following skills:

- ▸▸ Use a master slide to place objects and set styles
- ▸▸ Insert a shape onto a slide and change its colour
- ▸▸ Draw lines on the master slide
- ▸▸ Insert a clip art image on the master slide
- ▸▸ Insert information in a footer
- ▸▸ Include text on the slides including headings, subheadings, and bulleted lists
- ▸▸ Format text (i.e., font type (serif or sans serif), point size, text colour, text alignment, and enhancements (bold, italic, and underscore)

1 Load the presentation software Microsoft PowerPoint and create a new blank presentation.

Slide

2 Click on View and then on Master to bring up the slide master.

3 Click on the slightly larger first top slide in the list down the left-hand side of the screen. You will now see the following on your screen:

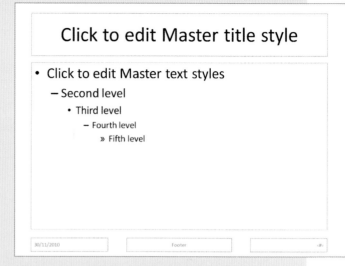

4 You are now going to place a pale grey rectangle that occupies about one quarter of the width of the slide. There are some placeholder boxes in the way of where the rectangle needs to go. You can simply click on them and drag and re-size them so that they are out of the way like this:

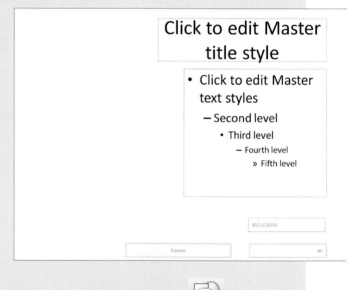

Click on Insert and then select Shapes and then choose the rectangle shape from the list of shapes by left clicking on it. You will now see the cursor change to a cross. Position this cursor on the top left-hand corner of the slide and, keeping the left mouse button pressed down, drag the cursor until you have drawn a rectangle like this:

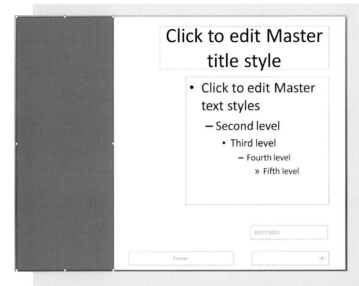

Notice the handles on the rectangle to show that it is still selected. You now need to change the colour of this rectangle to light grey.

Click on | Home | tab.

Click on 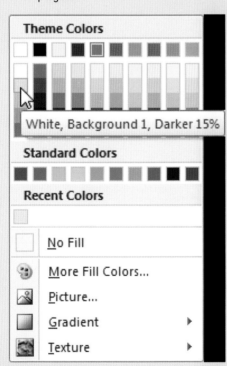 Shape Fill ▾ and the following colour palette is displayed:

Left click on the grey colour as shown above and the rectangular area on all the slides will be filled with this choice.

You are now going to put a red border around the rectangular area.

Click on ✎ Shape Outline ▾ and select the red colour from the standard colours by clicking on it.

The outline of the rectangular area is now in red.

The next step is to select the thickness of the border which is expressed in points. Click on

✎ Shape Outline ▾ again and this time select Weight from the menu. Now choose 1½ pt from the list of weights.

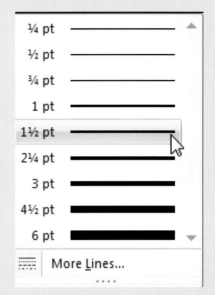

5 You are now going to insert two horizontal red lines near the top of the page.

Move the placeholder area out of the way to a position similar to that shown here:

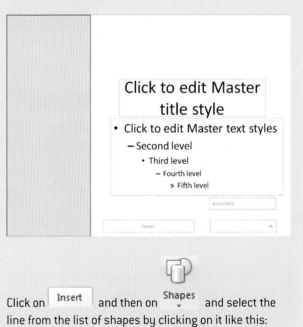

Click on | Insert | and then on Shapes ▾ and select the line from the list of shapes by clicking on it like this:

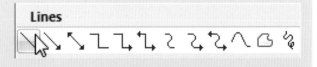

The cursor changes to the cross-wires. Position this cursor on the left at the start of the line and then drag the line until it meets the other side of the slide. If you press the shift key, it will keep the line in a horizontal position so you do not need to spend time trying to adjust the line yourself.

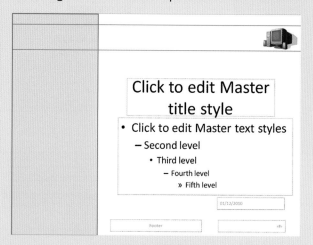

With the line still selected, click on **Format** and then click on ⬜ **Shape Outline ▾** and select red from the list of colours. Click on ⬜ **Shape Outline ▾** again and this time select Weight and then select a weight of 1½ pt.

The red line changes to the same width as that for the border of the rectangle. Repeat this by placing a line in red with weight 1.5 pt in the position shown here:

6 You are now going to insert a piece of clip art on the right of the slide and between the two red lines you have just inserted.

Click on **Insert** and then on **Clip Art** and then enter the search word "Computer" into the search box and click on Go. Then select an image of a computer (any will do).

Right click on the image and drag it into the approximate position. You will need to re-size the image by right clicking on a corner handle and dragging until it is of a size that will fit between the two lines. You will probably also have to move the image into the correct position.

Your image should now be in a position similar to this:

7 Move the footer to the left slightly so that it is nearer the red vertical line:

Click on **Insert** and then on **Header & Footer** when the header and footer dialogue box appears like this:

Click on the box for footer to indicate you want to include a footer.

Now enter the following footer details: your name, Centre number, and candidate number. If you do not have the Centre number use the number 1234 and if you do not have your candidate number, enter 56789.

Your dialogue box will now contain the footer details you entered:

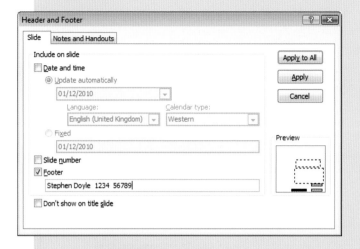

Click on Apply to All to apply the footer details to all the slides.

Note that only the footer details and not the date and slide number details will appear on the actual slides. This is because we have not selected to show the date and the slide number. Notice that the placeholder areas still appear on the master slide for these.

8 Your name, Centre number, and candidate number should be left aligned in black using a 12 point sans serif font. To do this, highlight the details in the footer like this:

Click on Home and then select any sans serif font (Arial has been chosen here) and set the size to 12 points like this:

Arial   12

Set the font colour to black by clicking on the font colour icon A⋅ and choosing black from the palette.

To align the text to the left, click on left align ≣ .

Check that the text in the footer appears like this:

Stephen Doyle 1234 56789

9 You are going to add the following text at the top of the master slide.

Compusolve

This text needs to be positioned at the top of the slide between the two red horizontal lines in a black bold 44 point sans serif font. The text also needs to be left aligned.

To put the text in this position you need to create a text box.

A

Click on Insert and then on Text Box . Now drag out a rectangular box in the position shown here:

Now type in the text Compusolve and highlight the text:

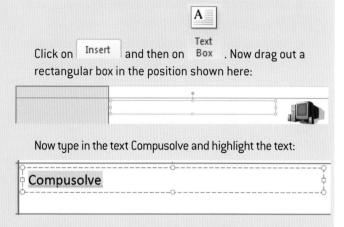

Click on Home and select a sans serif font such as Arial and a point size of 40 pt as shown here:

Arial   40

The text is already left aligned and we know this because the left align button is shown in orange (i.e., ≣).

Make the text bold by clicking on **B** .

Your text in the text box should now look like this:

Compusolve

10 You are now going to set the styles of text throughout the presentation. This will ensure that text used for headings, subheadings, and bulleted lists is consistent across all the slides used in the presentation.

Align the placeholder areas as shown here:

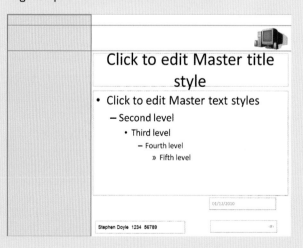

You have been asked to set the following styles of text throughout the presentation:

- ⏩ Heading: sans serif font, red, left aligned, and 40 point
- ⏩ Subheading: sans serif font, centre aligned, and 32 point
- ⏩ Bulleted list: sans serif font, black, left aligned, and 24 point. You can choose your own bullet.

Highlight the text in the "Master title style" placeholder area.

You are now going to format the text to that described for the heading.

Click on ⟨ Home ⟩ and then on change the font to a sans serif font such as Arial and change the size to 40 pt

The text is already centre aligned but you need to change the colour to red by clicking on ⟨**A**⟩ and choosing red from the palette of colours.

Notice that this slide does not have a subheading on it but it does contain a bulleted list, so we will change the bulleted list.

Highlight the text in the line containing the bullet and the text "Click to edit Mast text styles" like this:

• **Click to edit Master text styles**

Click on ⟨ Home ⟩ and then on change the font to a sans serif font such as Arial and change the size to 24 pt

Arial   ▾ 24 ▾

The text is already left aligned because the left align button is shown in orange (i.e., ⟨▤⟩).

To change the bullet to another shape click on the drop down arrow on the bullet icon ⟨▤▾⟩ and the following choice of bullets appears:

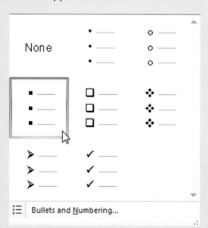

Click on the Filled Square Bullets to select that shape of bullet.

The bullet now changes to this:

▪ **Click to edit Master text styles**

11 You now have to change the subheading. To do this you need to have one of the slides that contain a subheading placeholder area displayed.

Choose the second slide in the list by clicking on it.

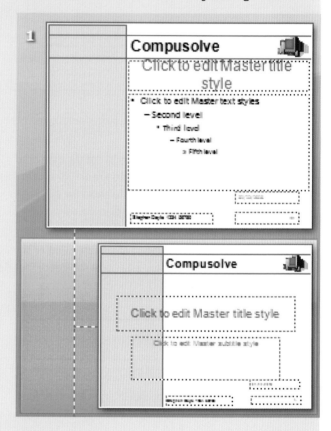

This slide appears like this:

You can see that some of the placeholder areas overlap the grey rectangle, so you need to move and re-size them so that they look similar to that shown below:

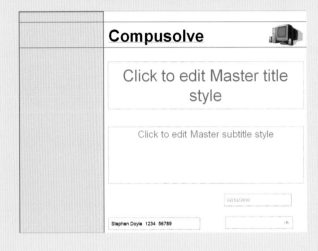

Highlight the text "Click to edit Master subtitle style".

Change the font if necessary to a sans serif font such as Arial and change the size to 32 points like this:

Centre the text by clicking on ☰ if the text is not already centred and change the font colour to black 🅰 ▾.

12 You have now completed the master slides which lay out the designs, but not the content for the slides in the presentation.

Save your work using the filename **Compusolve_v1**

Activity 16.3

Adding the content in the placeholder areas to create the slides

In this activity you will learn the following skills:

▸▸ Include text on the slides including: headings, subheadings, and bulleted lists

▸▸ Add text to a bulleted list in two columns

▸▸ Insert a text box

▸▸ Animate the addition of items in a bulleted list

▸▸ Create a chart within the presentation package

▸▸ Include segment labels, remove the legend, move and re-size the chart

▸▸ Add presenter notes to slides

1 Load the presentation file **Compusolve_v1** if it is not already loaded.

2 You are now going to add the content for the first slide. Here is what you have to do:

Enter the heading: Solving your computer problems

Enter the subheading: Hardware and software problems solved

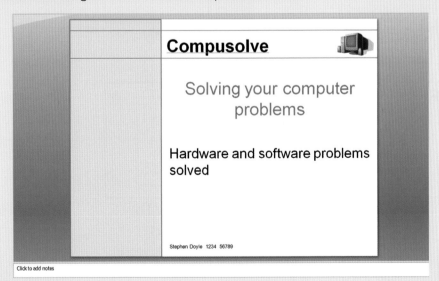

3 Notice the area below the slide containing the message "Click to add notes".

This is where you type in your presenter notes. These will appear on the presenter's computer screen but the audience will only see the slides.

In this area type in the following text:

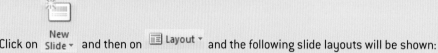

This brief presentation will introduce our business and outline the benefits of our unique service

4 You are now going to add the content to the second slide.

This slide is going to contain two bulleted lists – both in the white area of the slide with one list down the left-hand side and the other down the right-hand side.

You therefore need to pick a layout which includes this feature.

Click on New Slide ▾ and then on ▦ Layout ▾ and the following slide layouts will be shown:

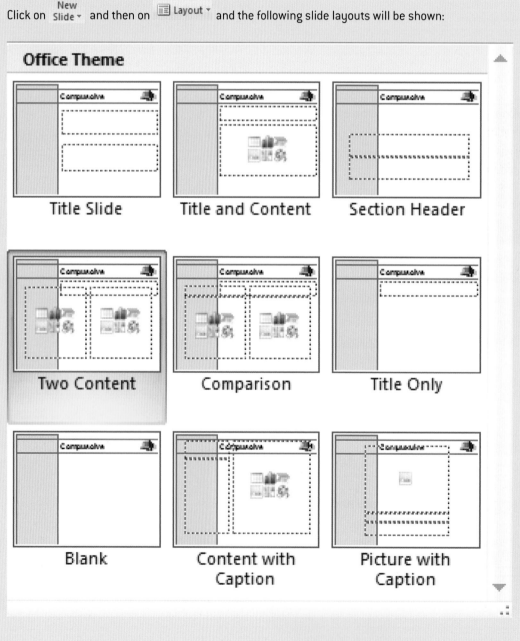

Choose the "Two content" layout by clicking on it, as this layout includes two bulleted lists side by:side.

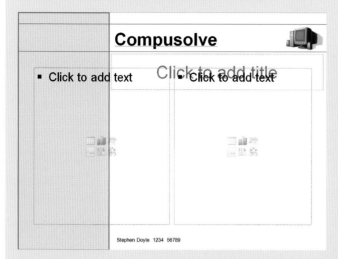

This slide contains overlapping placeholder areas.

Also, you have to put the subheading "Helping businesses and individuals" on the slide. There is no heading for this slide.

Click on the outline of the placeholder area for the heading (i.e., the box containing the text "Click to add title" and press the backspace key to delete it.

Your slide should now look like this:

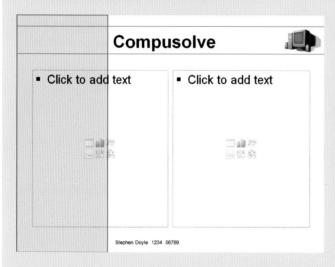

You now have to add a subtitle to the slide. This is to be positioned just above the two columns for the bullet points.

Before this is done you need to create some space and at the same time move the two placeholder areas for the bullets. When you have done this your slide should look like this:

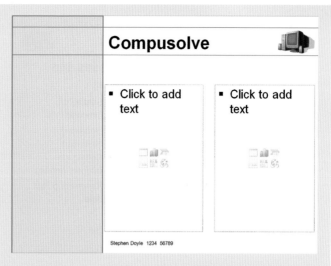

You now need to add a subtitle with the text "Helping businesses and individuals" in it.

There are several ways of doing this but the easiest is simply to create a text box and check that the text has the formatting of a subtitle.

Click on `Insert` and then on Text Box and then create a text box in the position shown here:

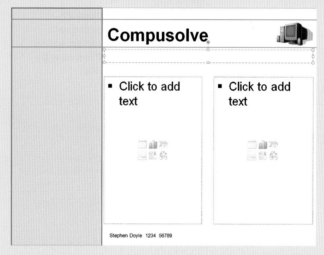

Now type in the following text for the subtitle: Helping businesses and individuals

Now format the text for the subheading to the text used for subheadings (i.e., subheading: sans serif font, centre aligned, and 32 point).

If the place holder areas for the bullets overlap with the subtitle, you will need to move the placeholder areas down slightly.

Now type in the bulleted text as shown here – check that the point size is 32 and if it is not this then you will need to change it.

Move your cursor over each of the animations in turn to see how they animate the bullet points on the slide.

Select the animation shown here:

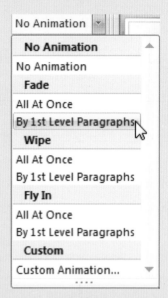

This will animate the bullet points by making each bullet appear gradually.

Animate the bullets in the second column in a similar way. For consistency make sure that you use the same type of animation for both columns of bullets.

Now add the following presenter notes:

Here are some of many services performed by our organization.

Compusolve

Helping businesses and individuals

- Installing hardware
- Copying files
- Installing software
- Removing viruses

- Setting up networks
- Repairing hardware
- Setting up Internet access
- Recovering lost data

Stephen Doyle 1234 56789

5 You are now going to animate the bullets so that they appear one at a time.

Select the first column of bullet points like this:

Compusolve

Helping businesses and individuals

- Installing hardware
- Copying files
- Installing software
- Removing viruses

- Setting up networks
- Repairing hardware
- Setting up Internet access
- Recovering lost data

Stephen Doyle 1234 56789

Click on ⫟ Animate: No Animation ▾ and the following animation choices appear:

6 You are now going to enter the content for the third slide.

Click on New Slide ▾ and then on ▦ Layout ▾ and the following slide layouts will be shown:

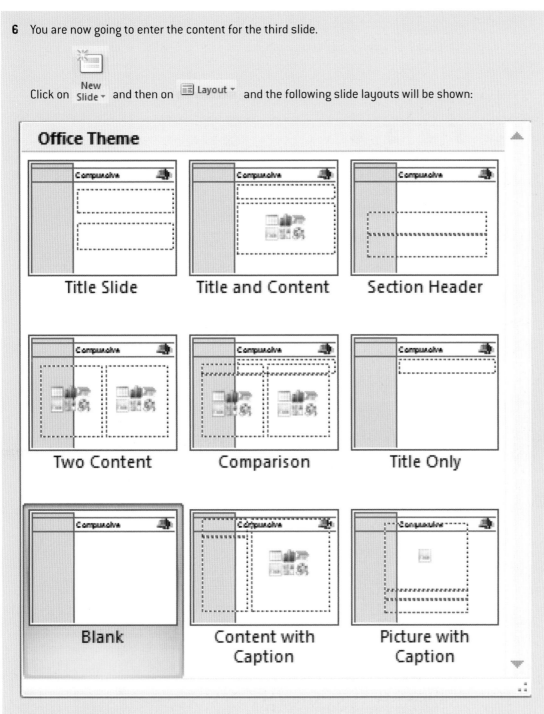

Click on Blank.

7 You are going to create a pie chart using the data in the following table on this slide:

Problem	% of calls
Virus infection	10
Hardware failure	28
Lost data	15
Lack of network	30
Other problems	17

Click on Insert and then on Chart and select ⬤ Pie from the list of charts.

Choose  and click on OK.

The following screen appears:

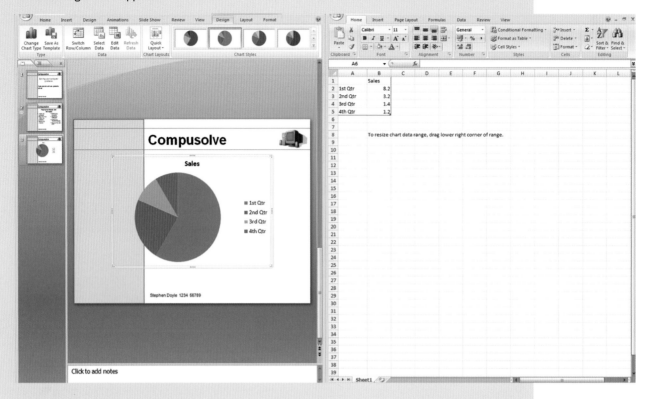

Notice that an Excel spreadsheet has appeared at the side of the slide. This spreadsheet is used to enter the data used to create the pie chart.

You can see that there is a table of data:

	A	B
1		Sales
2	1st Qtr	8.2
3	2nd Qtr	3.2
4	3rd Qtr	1.4
5	4th Qtr	1.2

You need to replace this data with your own data like this:

	A	B
1	**Problem**	**% of calls**
2	Virus infection	10
3	Hardware failure	28
4	Lost data	15
5	Lack of network	30
6	Other problems	17

You will then notice that the pie chart is produced automatically on the slide like this:

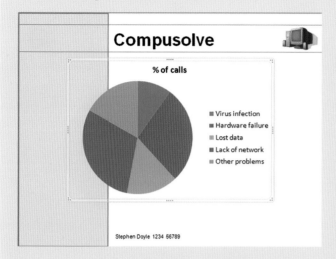

8 You are now required to put the percentages on the segments as well as the problem and then delete the legend.

Right click on one of the sectors of the pie chart and this menu appears:

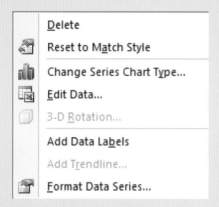

Click on Add Data Labels.

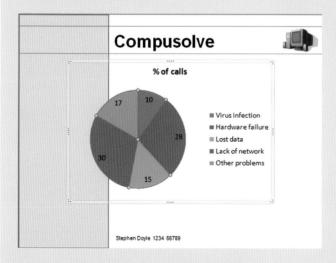

The numbers for the percentages are now added to the chart.

Right click on a segment in the chart again and select Format Data Labels.

The format labels dialogue box now appears like this:

9 You are required to show the value and the category names.

Apply these to the dialogue box like this:

Click on Close to apply the settings.

The chart now appears like this:

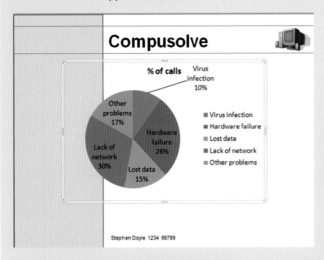

Notice that the text for "Virus infection 10%" has not appeared in the segment. This is because the text is too wide to fit in the segment.

10 You have been asked to remove the legend from the chart.

To remove the legend, click on it so that the handles appear like this:

Press the backspace key and the legend is removed.

Position the cursor on the corner of the chart border like this:

Now keep the right mouse button pressed down and drag the corner to make the pie chart larger. Also ensure that the pie chart is central in area.

Make the chart as large as possible to fit the white space like this:

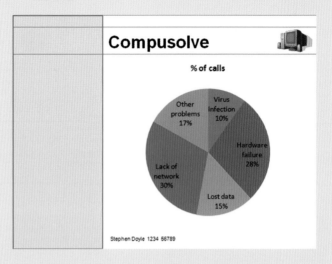

11 Now add the following presenter notes to complete this slide:

You can see from the chart that the majority of the calls for our services are about network problems, although calls about hardware failure are a close second.

12 Click on New Slide ▾ to start adding content on the fourth slide and then click on 🗔 Layout ▾ where you can choose the layout best suited to the slide you want to produce.

You are required to produce a slide with a subheading, a single bulleted list that is left aligned and a piece of clipart.

You can see from the list of slides that there is no one slide which meets all the requirements:

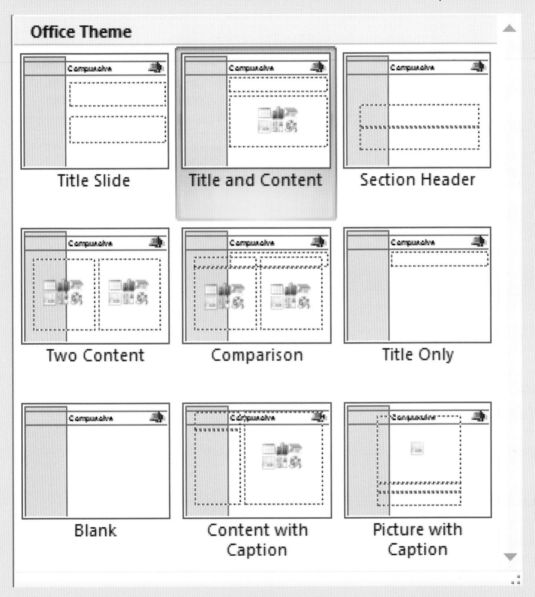

The most appropriate slide is the one called "Title and Content" but this has a heading (called a title) rather than a subheading but it is easy to change it.

Click on the "Title and Content" slide and the following slide appears:

Highlight the text "Click to add title" and make the following changes to the text so that it is formatted as a subheading:

Change the font if necessary to a sans serif font such as Arial and change the size to 32 points like this:

Arial ▼ 32 ▼ .

Centre the text by clicking on if the text is not already centred and change the font colour to black

A ▼ .

Now enter the following text for the subheading:

In addition to dealing with
computer problems we can:

Enter the following bulleted list on the left-hand side of the slide:

▸▸ Offer advice
▸▸ Provide training
▸▸ Provide backup services

Place a different clipart image of a person using a computer or a group of people using computers in the white section at the bottom of the bulleted list and in a similar position to that shown here:

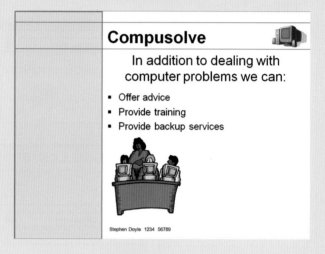

Now add these presenter notes to this slide:

Here are some of the other services that Compusolve can provide.

13 Save your presentation using the filename Compusolve_v2

Activity 16.4

Adding slide transitions, printing presenter notes, and providing screen shots for evidence
In this activity you will learn the following skills:

▸▸ Automate the transitions between slides
▸▸ Print a copy of the presentation slides including the presenter notes
▸▸ Print a copy of the audience notes
▸▸ Provide printed evidence of animation of bullet points
▸▸ Provide printed evidence of animation of slide transitions

1 Load the presentation software Microsoft PowerPoint and the file Compusolve_v2 if it is not already loaded.

2 You now have to use a transitional effect between each slide in the presentation.

3 Click on Animations and then move your cursor onto each of the slide transitions (shown below) in turn to see their effect.

Choose one of them (you choose which one) and then click on , which will apply the transition you have selected to all the slides in the presentation. Notice the following settings which can be changed:

Transition Sound: [No Sound]
Transition Speed: Fast
Apply To All

Advance Slide
☑ On Mouse Click
☐ Automatically After: 00:00

Notice that you can set the slide transitions to be performed automatically after a certain period of time if the presenter has not clicked on the mouse. You can also alter the speed of the transitions and also to make a sound during the transition.

4   Save the presentation using the filename:

Compusolve_final_version

5   A copy of the presentation slides needs to be printed which must show the presenter notes.

Click on and then choose Print and then Print Preview like this:

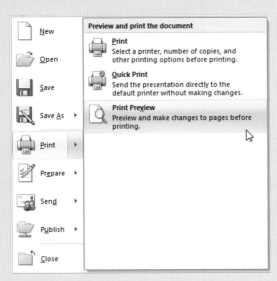

Now click on Slides and from the drop-down menu choose Notes Pages. You will now see each slide with the presenter notes on a separate page like this:

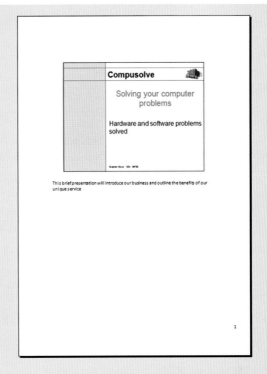

Click on Print to print all the pages.

The Print dialogue box appears similar to this:

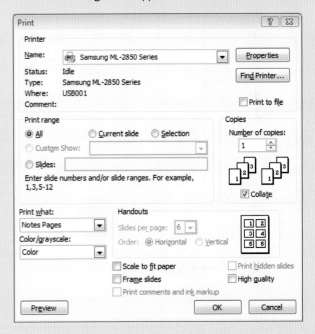

Notice all the settings which can be changed if needed here.

As all the settings are correct click on OK to print the slides.

These slides with presenter notes would usually be used by the presenter to refer to.

6   You now need to produce some printouts of the slides with space for the audience to make notes on each page. You have been asked to produce 3 slides on each page.

Click on the drop-down arrow in the

section and choose Handouts (3 Slides Per Page).

Slides
Handouts (1 Slide Per Page)
Handouts (2 Slides Per Page)
Handouts (3 Slides Per Page)
Handouts (4 Slides Per Page)
Handouts (6 Slides Per Page)
Handouts (9 Slides Per Page)
Notes Pages
Outline View

You will now see how the slides and note area will appear when printed:

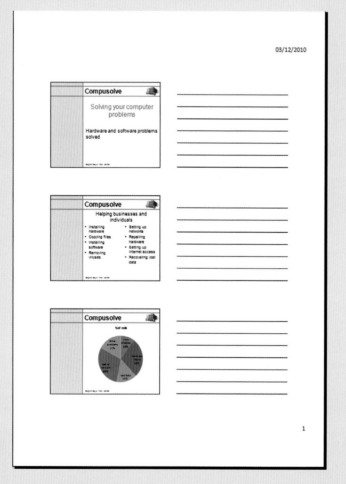

Print this document.

7   You now have to provide evidence that you have included the animations correctly.

Slide 2 contains the animations for the bulleted points to appear one at a time.

Click on this slide so that it is displayed like this:

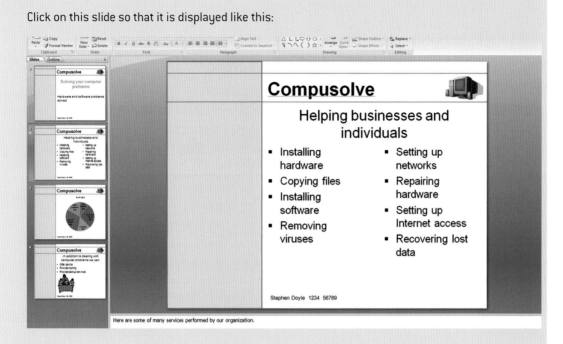

Click on [Animations] and then on [icon] Custom Animation .

The slide now appears with the bulleted points numbered like this:

The numbering 1, 2, 3, etc., indicates that each bullet is animated separately and it is the evidence that you have to supply. You now have to take a screenshot of this screen by pressing Prt Scr (i.e., Print Screen) key on your keyboard. The image of the screen is stored in the clipboard. You can then open a document using your word-processing software and then paste the image into a suitable position. You will need to make sure that you include your name, centre number, and candidate number on the document before printing it out.

8 You now have to produce evidence that the slide transitions have been applied according to the instructions in the examination.

Click on [Normal icon] if you are not looking at the normal slide view already.

Click on **Slide Sorter** and you will see the slides listed in the presentation appear like this:

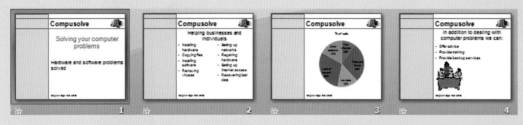

You can see the small symbol 📼☆ appear on each slide showing that a slide transition has been applied to each slide.

9 To obtain a printout of this evidence press the Prt Scr (i.e., Print Screen) button on your keyboard to copy it, and open a new document in your word-processing software. Paste the screenshot image in a suitable place and make sure that you add your name, centre number, and candidate number to the document which you should then print.

Manipulating images for use on slides

Sometimes in the examination you will be given an image that has to be manipulated in some way such as:

▶▶ Resizing (i.e., made bigger or smaller in order to fit a certain space on the slide)
▶▶ Positioning (i.e., the image needs to be selected and dragged into the correct position)
▶▶ Cropping (i.e., only using part of the image)
▶▶ Copying (i.e., so that the same image can be used in different places)
▶▶ Changing the image contrast (i.e., adjusting the difference between the light and dark parts of the image)
▶▶ Changing the brightness (i.e., making the whole image lighter or darker).

You can do all of the above using PowerPoint rather than use specialist image editing software.

Activity 16.5

Manipulating images

In this activity you will learn the following skills:

▶▶ Insert an image onto a slide
▶▶ Crop an image
▶▶ Resize an image
▶▶ Flip an image
▶▶ Adjust the contrast and brightness of an image

1 Load the presentation software Microsoft PowerPoint.

2 You need a blank slide on which to load an image.

   Click on 🗐 Layout ▾ and then from the list of layouts, select Blank.

➡

3 Click on Insert and then Picture — you now need to follow instructions from your teacher who will tell you where the image file called Caribbean can be found.

Once located, double left click on the image and it will be loaded onto the slide like this:

This image occupies the entire slide and there is also an unwanted object in the picture.

If the picture has not been selected (as shown by the small circles around the edges) then left click on the image to select it.

The picture formatting toolbar will automatically appear like this:

4 Click on the Crop tool in this toolbar .

Notice the handles around the edge of the picture. Click on the top right handle and drag it to the left. Use these handles to crop the picture so that the part containing the bottle is not used:

Click on the white part of the slide to deselect the cropping tool.

Click on the image again so that the handles are shown.

Click on the bottom right of the image and drag the corner until the image has a height of about half of the original like this:

5 You are now going to flip the image.

Select the image if it is not already selected.

Click on 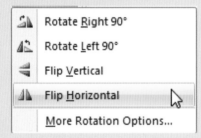 .

From the menu select Flip Horizontal:

⬆️	Rotate <u>R</u>ight 90°
⬆️	Rotate <u>L</u>eft 90°
◀	Flip <u>V</u>ertical
◀◀	Flip <u>H</u>orizontal
	More Rotation Options...

Notice the image has flipped with the boats in the image appearing on the right:

6 With the image selected move it to the top right-hand corner of the slide like this:

7 You now have to alter the contrast and the brightness of the image.

Ensure that the image is selected and the picture format tools are shown.

Click on ◑ Contrast ▾

Increase the contrast to +20%:

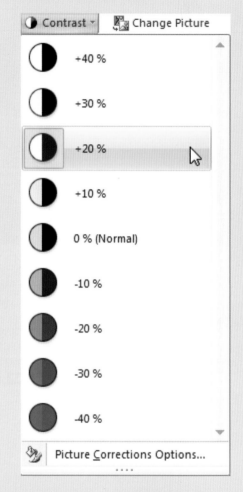

Click on ☀ Brightness ▾ and select +10%.

8 You now have to add a callout box containing text.

Click on | Insert | and then on Shapes ▾ and from the list of shapes click on the Rectangular callout box shown here:

You will see the cursor change to a cross. Click on a position near to the image and drag the cursor creating the callout box like this:

Click on the yellow diamond and drag onto the image in a similar position to that shown here:

With the callout selected, right click on it and choose Format Shape from the list.

The following Format Shape window appears:

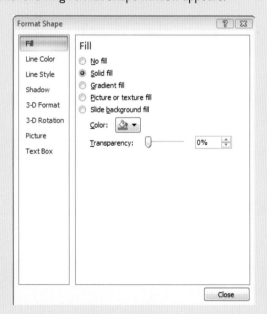

Notice Fill has been selected. Click on Color: and choose White.

Click on Line Color and click on Color: and choose Black.

Click on Close.

The callout box needs turning into a text box. Click on Text Box.

The cursor changes shape.

Click on Home and then click on the font colour A and choose the colour red. Change font to Arial and size to 10pt.

Click inside the callout and type in the following text:

Natural coves are used by ships to shelter from storms.

The callout box now appears on the slide like this:

9 In this step you will be putting a copyright symbol followed by the name (i.e., © S Doyle) underneath the image.

First create a text box by clicking on Text Box and position the box as shown here:

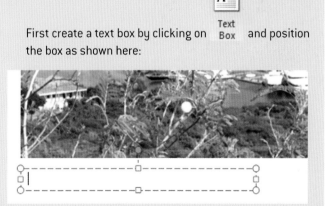

Change the font size to 10 pt.

Click on Insert and then on Symbol and click on the copyright symbol

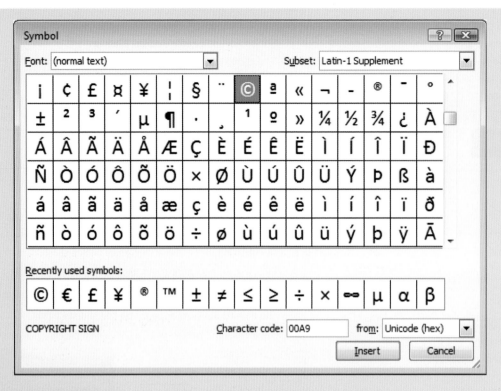

Click on [Insert] and the copyright symbol is inserted. Click on [Close].

Now add the text S Doyle

The text box should now look like this:

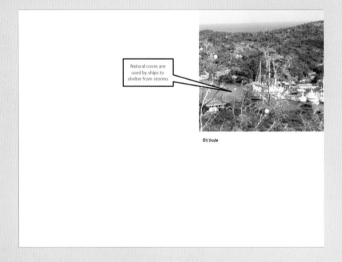

10 Save the slide using the filename **Caribbean**.

Abnormal data data that is unacceptable and that should be rejected by a validation check. For example, entering text into a numeric field or data which is outside the range specified.

Absolute reference a reference to a cell used in a formula where, when the formula is copied to a new address, the cell address does not change.

Access rights restrictions to a user's access to only those files they need in order to perform their job.

Actuator a device which takes control signals from a computer and converts them into movement, e.g. to turn a valve on/off or to open/close a window.

Address book the names and email addresses of all the people to whom you are likely to send email stored as a file.

Alphanumeric data sometimes called text and it includes letters, digits and punctuation marks.

Analogue a continuously changing quantity that needs to be converted to digital values before it can be processed by a computer.

Analogue-to-digital converter (ADC) a device that changes continuously changing quantities (such as temperature) into digital quantities.

Anti-virus software software that is used to detect and destroy computer viruses.

Application software software designed to do a particular job such as word-processing or database software.

Aspect ratio the ratio of the width of an image to its height.

Attachment a file which is attached to an email which the recipient can open and view the contents provided they have suitable software to open the file.

Backing storage storage which is not classed as ROM or RAM. It is used to hold programs and data. Backing storage devices include magnetic hard drives, optical drives (CD or DVD), flash/pen drives, etc.

Backup file copy of a file which is used in the event of the original file being corrupted (damaged).

Backup keeping copies of software and data so that the data can be recovered should there be corruption or loss of some or all of the ICT system.

Bandwidth a measure of the amount of data that can be transferred per second over the Internet or other network.

Bar code a series of dark and light lines of differing thickness which are used to represent a number which is usually written below the bar code. Can be read by an scanner to input the number accurately.

Batch processing type of processing where all the inputs needed are collected over a period of time and then batched together, inputted and processed in one go. For example, questionnaires from a survey are collected over a few weeks and then batched together and processed in one go.

bcc (blind carbon copy) this is useful when you want to send an email to one person and others but you do not want the others to see each other's email addresses.

Biometric a property of the human body such as fingerprints or pattern on the retina which can be used to identify a person and allow them access to a computer system.

BIOS (basic input/output system) stored in ROM and holds instructions used to 'boot' (i.e. start) the computer when first switched on.

Bit a binary digit 0 or 1.

Blog a web site that allows comments to be posted usually in reverse chronological order.

Blogger a person who maintains a blog.

Bluetooth a method used to transfer data over short distances from fixed and mobile devices wirelessly. The range of Bluetooth depends on

the power of the signal and can be typically from 5m to 100m.

Blu-ray optical disk that has a much higher storage capacity than a DVD. Blu-ray disks have capacities of 25 Gb, 50 Gb, and 100 Gb. These high capacity Blu-ray disks are used to store high definition video. They are used for storing films/movies with a 25 Gb Blu-ray disk being able to store 2 hours of HDTV or 13 hours of standard definition TV. It is possible to play back video on a Blu-ray disk whilst simultaneously recording HD video.

Bookmarks storage area where the URL (i.e. the web address) of a web site can be stored so that it can be accessed later using a link.

Boolean data data that can only exist in two states. For example True/False.

Bridge a hardware device used to connect two local area networks to each other. The purpose of a bridge is to decide whether a message needs to be transferred between the two networks or just confined to one of them. This reduces network traffic.

Browser (also called web browser) software program you use to access the Internet. Microsoft Internet Explorer is an example of a web browser.

Bullet point a block or paragraph of text that has a symbol placed in front to make the section of text stand out.

Bus topology type of network topology where all the computers are connected to a common shared cable called the bus.

CAD (computer-aided design) software software used to produce technical drawings, plans, designs, maps, etc.

cc (carbon copy) used when you want to send an email to one person but you also want others to see the email you are sending. To do this you enter the email address of the main person you are sending it to and in the box marked cc you enter all the email addresses, separated by commas, of all the people you wish to receive a copy.

CD R (compact disk recordable) optical storage where data is stored as an optical pattern. The user can record their data onto the disk once only.

CD ROM (compact disk read only memory) optical storage where data is stored as an optical pattern. Once data has been written onto CD ROM it cannot be erased. It is mainly used for the distribution of software.

CD RW (compact disk read-write) optical storage that allows data to be stored on the disk over and over again just like a hard disk. This is needed if the data stored on the disk needs to be updated. You can treat a CD RW like a hard drive but the transfer rate is less and the time taken to locate a file is greater. The media is not as robust as a hard drive.

Cell an area on a spreadsheet produced by the intersection of a column and a row in which data can be placed.

Changeover the process by which an older ICT system is replaced with a newer one.

Character any symbol (letter, number, punctuation mark, etc.) that you can type from the keyboard.

Check digit a decimal number (or alphanumeric character) added to a number for the purpose of detecting the sorts of errors humans normally make on data entry.

Chip and PIN chip readers are the devices into which you place a credit/debit card to read the data which is encrypted in the chip on the card. The PIN pad is the small numeric keypad where the personal identification number (PIN) is entered and the holder of the card can be verified as the true owner of the card.

Clipboard temporary storage area used for copying or cutting data to and then pasting it somewhere else.

Cloud computing Internet-based computing where programs and data are stored on the Internet rather than on the user's own computer.

Command line interface type of user interface where a user has to type instructions in a certain format to accomplish a task.

Compression storing data in a format that requires less space. A compressed file takes less time to be transferred across a network.

Computer Misuse Act the Act which makes illegal a number of activities such as deliberately planting viruses, hacking, using ICT equipment for frauds, etc.

Content the actual text, images, etc.

Control system system used to control a process automatically by making use of data from sensors as the input to the system.

Copyright, Designs and Patents Act an Act making it a criminal offence to copy or steal software or use the software in a way that is not allowed according to the software licence.

CPU (central processing unit) the computer's brain. It interprets and executes the commands given to it by the hardware and software.

Cropping only using part of an image.

CSV (comma separated variables) a way of holding data in a file so that it can be transferred into databases or spreadsheets.

Data raw facts and figures, e.g. readings from sensors, survey facts, etc.

Database a series of files stored in a computer which can be accessed in different ways.

Data capture term for the various methods by which data can be entered into the computer so that it can be processed.

Data logger a device which collects readings from one or more sensors. The time interval between each reading can be varied (called the logging rate) and the total time over which the data is logged (called the logging period) can also be varied.

Data logging the process of using an ICT system to collect data from sensors at a certain rate over a certain period of time. Remote weather stations use data logging.

Data Protection Act an Act that restricts the way personal information is stored and processed on a computer.

Data redundancy where the same data is stored more than once in a table or where the same data is stored in more than one table.

Data type check validation check to ensure the data being entered is the same type as the data type specified for the field.

Digital camera a camera that takes a picture and stores it digitally so that it can be transferred to and processed by a computer or other device.

Digital signature a way of ensuring that an email or document sent electronically is authentic. It can be used to detect a forged document.

Digital-to-analogue converter (DAC) a device that changes digital quantities into analogue ones.

Direct/random access data is accessed immediately from the storage media. This is the method used with storage media such as magnetic hard disks and optical media such as CD and DVD.

Dot matrix printer a printer which uses numerous tiny dots to make up each printed character. It works by hitting tiny pins against an inked ribbon to make the dots on the page, so this makes this printer noisy.

Double entry of data two people use the same data source to enter the details into the ICT system and only if the two sets of data are identical, will they be accepted for processing. It is a method of verification.

Download to copy files from a distant computer to the one you are working on.

Drag and drop allows you to select objects (icons, folders, files, etc.) and drag them so that you can perform certain operations on them such as drag to the recycle bin to discard, add a file to a folder, copy files to a folder and so on.

DVD R (DVD recordable) a type of optical storage. DVD R allows data to be stored on a DVD only once.

DVD ROM (digital versatile disk read only memory) DVD ROM is optical storage and offers much higher storage capacity compared to CD. It is used for the distribution of movies where you can only read the data off the disk. A DVD ROM drive can also be used for reading data off a CD. DVD is mainly used for the distribution of films and multimedia encyclopaedias.

EFTPOS (electronic funds transfer at point of sale) where electronic funds transfer takes place at a point of sale terminal. This means that money is transferred from the bank or credit card company to the store when you pay for goods at a store.

Encryption the process of scrambling files before they are sent over a network to protect them from hackers. Also the process of scrambling files stored on a computer/storage device so that if the computer/storage device is stolen, the files cannot be read. Only the person who has a special key can see the information in its original form.

EPOS (electronic point of sale) a computerized till which can be used for stock control.

Ergonomics an applied science concerned with designing and arranging things people use so that the people and things interact most efficiently and safely.

Evaluation the act of reviewing what has been achieved, how it was achieved and how well the solution works.

Expert system an ICT system that mimics the decision-making ability of a human expert.

Extreme data is data on the borderline of what the system will accept. For example, if a range check specifies that a number from 16 to 21 inclusive is entered, the extreme data would be 16 and 21.

Favourites storage area where the URL (i.e. the web address) of a web site can be stored so that it can be accessed later using a link.

Fax a machine capable of sending and receiving text and pictures along telephone lines.

Field a space in an information handling system/database used for inputting data. For instance, you could have fields for surname, date of birth, etc.

File a collection of related data.

File attachment (sometimes called an attachment) a file that is attached to an email and can be sent to another person or a group of people.

File compression taking files and using special software to reduce their size before sending them over the Internet or to reduce their size so that they take up less space on the storage media.

Firewall a piece of software, hardware or both that is able to protect a network from hackers.

Flash/pen drives portable storage media which offer cheap and large storage capacities and are an ideal media for photographs, music and other data files. They consist of printed circuit boards enclosed in a plastic case.

Flat file method used for storage of data in a database where all the data is held in a single table.

Font a set of letters and characters in a particular design.

Footer text placed at the bottom of a document.

Format checks checks performed on codes to make sure that they conform to the correct combinations of characters.

Gateway the device/software that translates between two different kinds of computer networks (e.g. between a WAN and a LAN).

GIF (Graphics Interchange Format) file type used for images. Images in this format are reduced to a maximum of 256 colours. Images in this format are compressed so this means that they load quickly. Used for simple line diagrams or clip art.

GIGO (garbage in garbage out) means that if you put rubbish into the computer then you get rubbish out.

GPS (Global Positioning System) system which uses the signals from several satellites to obtain the exact position

of any object (e.g. aircraft, ship, car, etc.) on the Earth's surface. Many cars are equipped with satellite navigation systems which use GPS so that the driver can locate their position on a map on a small screen inside the car.

Graphics tablet an input device which consists of shapes and commands on a tablet which can be selected by the user by touching. Using a graphics tablet, means that more space is left on the screen for a plan or diagram.

Graph plotter output device which draws by moving a pen. Useful for scale drawings and is used mainly with CAD packages.

GUI (graphical user interface) interface that allows users to communicate with the computer using icons and pull-down menus.

Hackers people who try (succeed) to break into a computer/computer network illegally.

Hacking process of trying to break into a secure computer system.

Hard copy printed output from a computer which may be taken away and studied.

Hardware the physical components of a computer system.

Header text placed at the top of a document.

Hot spot an image or piece of text used as a link. When you click on the image or text, you are taken to another part of the same page, a different page or a different site, or it may open a new file or a new window.

Hub a hub contains multiple ports (i.e. connection points). When a packet of data arrives at one port, it is copied to the other ports so that all network devices of the LAN can see all packets. Every device on the network will receive the packet of data, which it will inspect to see if it is relevant or not.

Hyperlink a feature of a web site that allows a user to jump to another web page, to jump to part of the same web page or to send an email message.

Hypertext Mark-Up Language (HTML) a computer programming language used to create documents on the

World Wide Web. You use it to specify the structure and layout of a web document.

Identity theft using your banking/credit card/personal details in order to commit fraud.

Inference engine one of three parts of an expert system. It is a rules base and is the part of the expert system that does the reasoning by manipulating and using the knowledge in the knowledge base.

Information Data + Meaning = Information. For example 12/03/11 is data. Only when we understand it is the 'date for the exam' and it is in dd/mm/yy do we have the information that the exam date is 12th March 2011. So date needs 'field name' and 'format' to become information.

Inkjet printer printer that works by spraying ink through nozzles onto the paper.

Input device the hardware device used to feed the input data into an ICT system such as a keyboard or a scanner.

Instant messaging (IM) a method of two people using real time text to conduct a conversation using the Internet.

Integer a whole number which can be positive, negative or zero.

Interactive where there is a constant dialogue between the user and the computer.

Internet a huge group of networks joined together. The largest network in the world.

Internet service provider (ISP) a company that provides users with an Internet connection.

Intranet a private network used with an organization that makes uses of Internet technology used for sharing internal information.

IP (Internet Protocol) address a number which uniquely indentifies the physical computer linked to the Internet.

Joystick input device used instead of the cursor keys or mouse as a way of producing movement on the screen.

JPEG a file format used for still images which uses millions of colours and compression which makes it an ideal file format for photographic images on web pages.

K Kilobyte or 1024 bytes. Often abbreviated as Kb. A measure of the storage capacity of disks and memory.

Key field this is a field that is unique for a particular record in a database.

Knowledge base one of three parts of an expert system. A huge organized set of knowledge about a particular subject. It contains facts and also judgemental knowledge, which gives it the ability to make a good guess, like a human expert.

LAN (local area network) a network of computers on one site.

Landscape page orientation where the width is greater than the height.

Laser printer printer which uses a laser beam to form characters on the paper.

Length check validation check to make sure that the data being entered has the correct number of characters in it.

Light pen input device used to draw directly on a computer screen or used to make selections on the screen.

Login accessing an ICT system usually by entering a user-ID/username or a password.

Magnetic stripe stripe on a plastic card where data is encoded in a magnetic pattern on the stripe and can be read by swiping the card using a magnetic stripe reader.

Magnetic stripe reader hardware device that reads the data contained in magnetic stripes such as those on the back of credit cards.

Mail merge combining a list of names and addresses with a standard letter so that a series of letters is produced with each letter being addressed to a different person.

Main internal memory memory which is either ROM (read only memory) or RAM (random access memory).

Master slides (also called slide masters) used to help ensure consistency from slide to slide in a presentation. They are also used to place objects and set styles on each slide. Using master slides you can format titles, backgrounds, colour schemes, dates, slide numbers, etc.

Megabyte (Mb) a unit of file or memory size that is 1024 kilobytes.

Megapixel one million pixels (i.e. dots of light).

Memory cards thin cards you see in digital cameras used to store photographs and can be used for other data.

Memory stick/flash drive/pen drive solid state memory used for backup and is usually connected to the computer using a USB port.

Microprocessor the brain of the computer consisting of millions of tiny circuits on a silicon chip. It processes the input data to produce information.

MIDI (musical instrument digital interface) used mainly to communicate between electronic keyboards, synthesizers and computers. MIDI files are compressed and the files are quite small.

Monitor another name for a VDU or computer screen.

MP3 music file format that uses compression to reduce the file size considerably which is why the MP3 file format is popular with portable music players such as iPods and mobile phones.

Multifunction devices hardware which brings together the functions of several devices. For example, by combining PDA and mobile phone technology you can have phones capable of browsing the Internet.

Multimedia making use of many media such as text, image, sound, animation and video.

Multimedia projector output device used to project the screen display from a computer onto a much larger screen that can be viewed by a large audience.

Network a group of ICT devices (computers, printers, scanners, etc.) which are able to communicate with each other.

Networking software systems software which allows computers connected together to function as a network.

Normal data data that is acceptable for processing and will pass the validation checks.

OCR (optical character recognition) a combination of software and a scanner which is able to read characters into the computer.

OMR (optical mark reader/ recognition) reader that detects marks on a piece of paper. Shaded areas are detected and the computer can understand the information contained in them.

Online processing the system is automatically updated when a change (called a transaction) is made. This means that the system always contains up-to-date information. Online processing is used with booking systems to ensure seats are not double booked.

Online shopping is shopping over the Internet, as opposed to using traditional methods such as buying goods or services from shops or trading using the telephone.

Operating system software software that controls the hardware of a computer and used to run the applications software. Operating systems control the handling of input, output, interrupts, etc.

Optical character recognition (OCR) input method using a scanner as the input device along with special software which looks at the shape of each character so that it can be recognized separately.

Optical disk a plastic disk used for removable storage includes CD and DVD.

Optical mark recognition (OMR) the process of reading marks (usually shaded boxes) made on a specially prepared document. The marks are read using an optical mark reader.

Output the results from processing data.

Password a series of characters chosen by the user that are used to check the identity of the user when they require access to an ICT system.

PDA (personal digital assistant) a small hand-held computer.

Personal data data about a living identifiable person which is specific to that person.

Pharming malicious programming code is stored on a computer. Any users who try to access a web site which has been stored on the computer will be re-directed automatically by the malicious code to a bogus web site and not the web site they wanted. The fake or bogus web site is often used to obtain passwords or banking details so that these can be used fraudulently.

Phishing fraudulently trying to get people to reveal usernames, passwords, credit card details, account numbers, etc., by pretending they are from a bank, building society, or credit card company, etc. Emails are sent asking recipients to reveal their details.

PIN (personal identification number) secret number that needs to be keyed in to gain access to an ATM or to pay for goods/services using a credit/debit card.

Piracy the process of illegally copying software.

Pixel a single point in a graphics element or the smallest dot of light that can appear on a computer screen.

Plotter a device which draws by moving a pen, and is useful for printing scale drawings, designs and maps.

Podcast a digital radio broadcast created using a microphone, computer and audio editing software. The resulting file is saved in MP3 format and then uploaded onto an Internet server. It can then be downloaded using a facility called RSS onto an MP3 player for storing and then listening.

Point a length which is 1/72 inch. Font size is measured in points. For example, font size of 12pts means 12/72=1/6 inch which is the height the characters will be.

Portrait page orientation where the height is greater than the width.

Presence checks validation checks used to ensure that data has been entered into a field.

Print preview feature that comes with most software used to produce documents. It allows users to view the page or pages of a document to see exactly how they will be printed. If necessary, the documents can be corrected. Print preview saves paper and ink.

Process any operation that transfers data into information.

Processing performing something on the input data such as performing calculations, making decisions or arranging the data into a meaningful order.

Processor often called the CPU and is the brain of the computer consisting of millions of tiny circuits on a silicon chip. It processes the input data to produce information.

Program the set of step-by-step instructions that tell the computer hardware what to do.

Proof-reading carefully reading what has been typed in and comparing it with what is on the data source (order forms, application forms, invoices, etc.) for any errors or just reading what has been typed in to check that it makes sense and contains no errors.

Proxy-server a server which can be hardware or software that takes requests from users for access to other servers and either forwards them onto the other servers or denies access to the servers.

Query a request for specific information from a database.

RAM (random access memory) type of main internal memory on a chip which is temporary/volatile because it loses its contents when the power is removed. It is used to hold the operating system and the software currently in use and the files being currently worked on. The contents of RAM are constantly changing.

Range check data validation technique which checks that the data input to a computer is within a certain range.

Read only a user can only read the contents of the file. They cannot alter or delete the data.

Read/write a user can read the data held in the file and can alter the data.

Real-time a real-time system accepts data and processes it immediately. The results have a direct effect on the next set of available data.

Real-time processing type of processing where data received by the system is processed immediately without any delay and is used mainly for control systems, e.g. autopilot systems in aircraft.

Record the information about an item or person. A row in a table.

Relational database a database where the data is held in two or more tables with relationships (links) established between them. The software is used to set up and hold the data as well as to extract and manipulate the stored data.

Relative reference when a cell is used in a formula and the formula is copied to a new address, the cell address changes to take account of the formula's new position.

Report the output from a database in which the results are presented in a way that is controlled by the user.

Resolution the sharpness or clarity of an image.

ROM (read only memory) type of internal memory on a chip which is permanent/non-volatile and cannot have its contents changed by the user. It is used to hold the boot routines used to start the computer when the power is switched on.

Router hardware device which is able to make the decision about the path that an individual packet of data should take so that it arrives in the shortest possible time. It is used to enable several computers to share the same connection to the Internet.

RSI (repetitive strain injury) a painful muscular condition caused by repeatedly using certain muscles in the same way.

rtf (rich text format) file format that saves text with a limited amount of formatting. Rich text format files use the file extension '.rtf'.

Sans serif a set of typefaces or fonts that do not use the small lines at the end of characters which are called serifs.

Scanner input device that can be used to capture an image and is useful for digitizing old non-digital photographs, paper documents or pictures in books.

Screenshot copy of what is seen on a computer screen. It can be obtained by pressing the Prt Scr button when a copy of the screen will be placed in the clipboard. The copy of the screen can then be pasted.

Search engine program which searches for required information on the Internet.

Secondary memory storage other than ROM or RAM and is non-volatile which means it holds its contents when the power is removed. It is used to hold software/files not being used.

Sensors devices which measure physical quantities such as temperature, pressure, humidity, etc.

Serial/sequential access data is accessed from the storage media by starting at the beginning of the media until the required data is found. It is the type of access used with magnetic tape and it is a very slow form of access when looking for particular data on a tape.

Serif a small decorative line added to the basic form of a character (letter, number, punctuation mark, etc.).

Social networking site a web site used to communicate with friends, family and to make new friends and contacts.

Software the actual programs consisting of instructions that allow the hardware to do a useful job.

Software licence document (digital or paper) which sets out the terms by which the software can be used. It will refer to the number of computers on which it can be run simultaneously.

Solid state backing storage the smallest form of memory and is used as removable storage. Because there are no moving parts and no removable media to damage, this type of storage is very robust. The data stored on solid state backing storage is rewritable and does not need electricity to keep the data. Solid state backing storage includes memory sticks/pen drives and flash memory cards.

Spam unsolicited bulk email (i.e., email from people you do not know, sent to everyone in the hope that a small percentage may purchase the goods or services on offer).

Spellchecker program usually found with a word-processor and most packages which make use of text which checks the spelling in a document and suggests correctly spelt words.

Spyware software that is put onto a computer without the owner's knowledge and consent with the purpose of monitoring the user's use of the Internet. For example, it can invade their privacy or monitor keystrokes, so it can be used to record usernames and passwords. This information can then be used to commit fraud.

Stand-alone computer if a computer is used on its own without any connection (wireless or wired) to a network (including the Internet), then it is a stand-alone computer.

Storage media the collective name for the different types of storage materials such as DVD, magnetic hard disk, solid state memory card, etc.

Style sheet a document which sets out fonts and font sizes for headings and subheadings, etc., in a document. Changes to a heading need only made in the style sheet and all the changes to headings in the document will be made automatically.

Swipe card plastic card containing data stored in a magnetic stripe on the card.

Switch a device that is able to inspect packets of data so that they are forwarded appropriately to the correct computer. Because a switch only sends a packet of data to the computer it is intended for, it reduces the amount of data on the network, thus speeding the network up.

Tags special markers used in HTML to tell the computer what to do with the text. A tag is needed at the start and end of the block of text to which the tag applies.

Tape magnetic media used to store data.

Templates electronic files which hold standardized document layouts.

Terabyte (Tb) a unit of file or memory size that is 1024 gigabytes.

Thesaurus software which suggests words with similar meanings to the word highlighted in a document.

TFT (thin film transistor) a thin screen used in laptops/notebooks or in desktops where desk space is limited.

Topology the way a particular network is arranged. Examples include ring, star, tree and bus.

Touchscreen a special type of screen that is sensitive to touch. A selection is made from a menu on the screen by touching part of it.

Tracker ball an input device which is rather like an upside down mouse and is ideal for children or disabled people who find it hard to move a mouse.

Transaction a piece of business, e.g. an order, purchase, return, delivery, transfer of money, etc.

Transcription error error made when typing data in using a document as the source of the data.

Transposition error error made when characters are swapped around so they are in the wrong order.

Tree topology a combination of two network topologies combined together, e.g. a series of star networks are connected onto a bus.

txt text files just contain text without any formatting. Text files use the file extension '.txt'.

Update the process of changing information in a file that has become out of date.

URL (Uniform Resource Locator) a web address.

USB (Universal Serial Bus) a socket which is used to connect devices to the computer such as web cams, flash drives, portable hard disks, etc.

User a person who uses a computer.

User interface the user interface uses an interactive screen (which can be a touchscreen) to present questions and information to the operator and also receives answers from the operator. Can also be one of the three parts of an expert system.

User log a record of the successful and failed logins and also the resources used by those users who have access to network resources.

Username or User-ID a name or number that is used to identify a certain user of the network or system.

Validation checks checks a developer of a solution sets/creates, using the software, in order to restrict the data that a user can enter so as to reduce errors.

Verification checking that the data being entered into the ICT system perfectly matches the source of the data.

Videoconferencing ICT system that allows virtual face-to-face meetings to be conducted without the participants being in the same room or even the same geographical area.

Virus a program that copies itself automatically and can cause damage to data or cause the computer to run slowly.

Voice recognition the ability of a computer to 'understand' spoken words by comparing them with stored data.

WAN (wide area network) a network where the terminals/computers are remote from each other and telecommunications are used to communicate between them.

Web cam a digital video camera that is used to capture moving images and is connected to the Internet so the video can be seen by others remotely. They are often included as part of the screen in computers or bought separately and connected to a USB port.

Web logs (blogs) blogs are web sites that are created by an individual with information about events in their life, videos, photographs, etc.

Web page a document that can be accessed using browser software.

Web site a collection of interconnected web pages relating to a topic or organization.

Wi-Fi a trademark for the certification of products that meet certain standards for transmitting data over wireless networks.

WIMP (Windows Icons Menus Pointing devices) the graphical user interface (GUI) way of using a computer rather than typing in commands at the command line.

WLAN A local area network (LAN) where some or all of the links are wireless, making use instead of infra-red or microwaves as a carrier for the data rather than wires.

World Wide Web (WWW) the way of accessing the information on all the networked computers which make up the Internet. WWW makes use of web pages and web browsers to store and access the information.

Index